AF225050

BAOFENG RADIO

Guerrilla's Guide

2024 ILLUSTRATED GUIDE

PATRICK VINCENT

Unlock Expert Skills to be Prepared in Any Emergency

Protect Yourself & The People You Love in Crisis Situations

Patrick has been fascinated by radio since he was a young boy. He would stay up late into the night, scanning through the bands on his parents' old handheld. Something was enchanting about hearing distant stations and trying to pull in faint signals from far-off places. The idea of voices broadcasting through the air, waiting for anyone with the right equipment to catch them, was simply magical to him.

As Patrick grew older, his curiosity about radio technology never waned. Whenever he had the opportunity, he would study these devices, especially walkie-talkie radios. When Patrick first got his hands on a Baofeng UV-5R, he quickly realized this budget-friendly little radio was hiding way more under the hood than the price tag let on.

For years, he has been a radio operator and has used Baofeng models extensively; it dawned on him that most people just used these things as glorified walkie-talkies. They had no clue about the potential waiting to be unleashed with a bit of know-how. That's the moment he knew he needed to write this guide to help other Baofeng users tap into all the juicy features and functions most people don't even know exist.

Drawing from his extensive experience using these radios in the field and teaching novices how to master them, Patrick set out to simplify all the technical aspects of radio waves, frequencies, and programming into plain English that anyone could understand—no confusing jargon, just crystal-clear step-by-step instructions to get you up and running like a pro.

Consider this book your backstage pass to the whole Baofeng experience. Patrick aims to directly put everything he's learned about these radios from years of experience into your brain. By the last page, you'll be a master at making even the cheapest Baofeng model work efficiently, whether that means boosting your range on remote hikes, keeping tabs on emergency channels in gnarly weather, or even cobbling together your own off-grid comms network when the internet goes out.

With the tricks and tips packed into this guide, that unassuming little Baofeng in your pocket will transform from a simple walkie-talkie into a core piece of your radio preparedness toolkit, ready to rock in any scenario life throws.

So, what is everyone waiting for? It's time to start this party and make some severe radio magic together, just like Patrick has been doing with countless other Baofeng enthusiasts in the real world!

Table Of Contents

INTRODUCTION

Many people get curious about those popular little walkie-talkie-style radios called "Baofengs" but don't know where to start learning the ropes. These affordable little Chinese gadgets seem like magical devices when you open the box, seeing a tangle of wires and intimidating settings. But once you decode some of the basics, a world of fantastic possibilities opens up.

I created this detailed guidebook to clue you in on the constructive tips and tricks for using Baofeng two-way radios effectively, even if you've never picked up a radio. No need to be a genius to track along with my straightforward explanations and real-world examples in plain English anyone could understand. I'll walk you from beginner confusion to expertly configuring secure local communications for your neighborhood, recreation, or volunteer activities in no time.

You'll learn key radio terms and technology breakdowns in a simple "here's what it means" style instead of just throwing dense generic definitions at you from Wikipedia, as some guides do. I cover everything from turning the radios on/off, replacing the stubby little antennas with upgraded models, and modifying secret privacy settings so outsiders can't listen in. Consider me like your new best friend guiding you causally rather than some stern professor lecturing technical nonsense to make radio operation seem intimidating.

This guidebook eliminates the problematic jargon and confusion. You'll gain fundamental skills for coordinating friends and family across impressive distances during outdoor adventures or uniting volunteer teams with clear communication. Most importantly, you'll be prepared to provide vital connectivity, coordinating regional help through proven radio tactics if local disasters knock out cellular networks and landlines when emergency responsiveness matters most.

So, what do you say we start unlocking the surprisingly great potential of these palm-sized black boxes together using the simple tips in my friendly guide? Just flip the page, and we'll decode the basics in no time...I promise you'll enjoy every chapter. Let the radio adventures begin.

CHAPTER 1: GETTING STARTED WITH BAOFENG RADIOS

1.1 Overview of Baofeng Radios

At first glance, Baofeng radios look like basic walkie-talkies—compact handheld devices with an antenna and some buttons. But they pack these impressive capabilities into a super affordable package! These little Chinese-made radios have earned a reputation for versatility and customization.

In a nutshell, Baofeng radios are radio transceivers, meaning they both receive and transmit radio signals in specific frequency bands. Unlike the average person's mental image of giant stationary radio towers, these are portable devices for easy use in vehicles, at remote work sites, while hiking, or carried discreetly in a pocket.

Baofeng models communicate in the VHF and UHF ranges. VHF covers high frequencies from 136 to 174 MHz without getting too technical. This includes bands reserved for commercial FM radio, marine, and other mobile services. UHF includes ultra-high frequencies from 400 to 520 MHz - the sweet spot for personal and community walkie-talkie operations on designated business/ amateur bands.

These radios can transmit and receive signals on select VHF and UHF frequencies, allowing two-way conversations across reasonable 2-5 miles (longer if you have a line of sight from high elevations). The standard "talk and listen" uses what most folks expect from traditional walkie-talkies or handheld transceivers.

But modern Baofeng radios pack a lot more versatility into the mix...

With the right app, they can receive (but not transmit on) regular FM radio stations like your car stereo or phone. This is handy for catching the occasional music or talk show in remote areas!

Many models also include a weather band receiver for tuning into those constant weather service broadcasts - no mobile app is required.

With some custom programming, you can set up separate frequency bands for different groups in your family, organization, or company. The radio will even scan through those channels and stop automatically when it detects activity.

There are also various configuration options, like setting a radio to unmute only when receiving a matching sub-code that follows your organization's customized standard. This gives a much-needed layer of privacy in crowded bands.

This guide will discuss specifics later regarding popular models, choosing the right one, programming, and critical features. But in short, these radios have earned a reputation for versatility, capability, and customization far beyond a traditional walkie-talkie. And all for an incredibly affordable price by tapping into Chinese manufacturing capabilities.

This comprehensive, easy-to-follow guidebook empowers readers of all experience levels with everything needed to fully utilize these versatile Baofeng two-way radios. It covers choosing the suitable Baofeng model, initial setup, programming frequencies, and channels, using advanced features like VOX and scanning, troubleshooting common issues, customization for privacy, and accessories to extend functionality. Whether you are a beginner or an expert radio user, this guide leaves no stone unturned - providing key insights and step-by-step instructions to help you master your Baofeng radio. From initial unboxing to advanced real-world operation, it gives you the knowledge to get the most out of these competent communication devices across various applications.

Key features found in many Baofeng radios include:

- ◆ Dual band monitoring (VHF 136-174 MHz, UHF 400-520 MHz).
- ◆ FM radio reception (65-108 MHz).
- ◆ Programmable channel memory.
- ◆ CTCSS and DCS coding for group communications.
- ◆ Battery saving functions.
- ◆ Transmit power from 1 to 8 watts.
- ◆ 128 programmable channels.
- ◆ Scanning capabilities.
- ◆ LED flashlight.
- ◆ Alarm functions -Cloning via cable.

With the proper programming and setup, these radios can be very effective communication tools for activities like hiking, camping, road trips, volunteer operations, and small business use. Their simplicity, performance, and low cost set Baofeng apart in the short-range two-way radio market.

1.2 Popular Baofeng Models

There is a wide variety of Baofeng radio models available. The most popular models purchased in the US include the UV-5R, BF-F8HP, UV-82, UV-B5, and UV-9R Plus models.

Baofeng UV-5R: This is the classic, most popular Baofeng model, known for its low cost and ease of programming. It is a dual-band transceiver, transmitting from 136-174 mHz VHF frequencies and 400-520 MHz UHF frequencies. Features include 128 programmable channels, FM radio reception, and customizable CTCSS and DCS operation.

Baofeng BF-F8HP: This upgraded version of UV-5R adds a high-performance receiver and enhanced transmit clarity. It also boasts more transmitter output power (8 watts vs 5 watts).

Baofeng UV-82: This is a more rugged, durable version of the UV-5R designed for outdoor recreation. It features improved water resistance and grip but provides VHF/UHF coverage with wide-band reception.

Baofeng UV-B5: This is the most affordable, stripped-down Baofeng model. It covers only the UHF 400-470 MHz range but maintains essential two-way radio functions. It is ideal for simple business/personal communications.

Baofeng UV-9R Plus: This is a higher-power (8-watt) dual-band model with physical buttons instead of menu-based controls for more straightforward operation. It also adds an extra 1800 mAh battery for 50% more life than standard models.

As you can see, there is quite a variety between models regarding frequency ranges, maximum output power, battery capacity, water resistance ratings, and ease-of-use features.

The UV-5R remains the number one seller - it strikes a fantastic balance between price, performance, and capability. However, buyers should evaluate their actual needs carefully.

If long battery life is critical for multi-day camping trips, spring for the UV-9R Plus model has a huge battery capacity!

If simple communication for site operations is needed across a large car dealership, nursing home, or hotel property, the UV-B5 configured for UHF channels will only save you a few bucks.

There's no need to over-buy on features you won't need or pay the premium for marine waterproofing if staying high and dry in the Rockies!

Hopefully, the quick snapshot of these popular models guides you as you evaluate options. There are even more obscure Baofeng radios, but these five models cover 90% of consumer and business use cases.

1.3 Choosing the Right Model

Determining the suitable Baofeng radio model involves prioritizing key factors like frequency coverage, transmit power, durability, battery life, programmability, and budget.

Frequency Coverage

Consider if you only need access to the 400-520MHz UHF spectrum for short-range personal communications. Or if you also need the 136-174MHz VHF frequencies for increased range capabilities. Some models offer broad VHF/UHF coverage, while more affordable options are UHF only.

Transmit Power

Maximum power output impacts the clarity and range of broadcasts. Typical handhelds emit anywhere from 1 watts to 8 watts. Higher wattage translates into farther-reaching signals but faster battery drain.

Physical Construction

Ruggedness is critical if the radio will be exposed to harsh weather, dropped often, or used in demanding environments. Look for IP55 ratings or mil-spec 810 construction to withstand dust, shock, and vibration.

Battery Capacity

The milliamp-hour (mAh) rating determines battery life. The standard capacity is around 1800 mAh, but higher-density 18650 packs up to 5000 mAh are available. Frequent users require substantially higher capacities.

Programming

Consider both current channel needs and future growth. Some offer simple manual programming, while computer software opens more advanced options. However, only a little can overwhelm inexperienced users.

Affordability

Avoid overbuying features that won't get utilized. But also don't sacrifice key capabilities to save a few dollars. Weigh price against functionality carefully when the budget is critical.

The "right" radio balances capability against cost based on expected usage scenarios. Carefully consider all factors before deciding which Baofeng model fits your needs.

Let's explore a few examples:

Little League Coaches

The UV-5R or UV-B5 models offer plenty for essential communications across a playing field complex and practice areas. They offer easy manual programming, 4-5 watt power, and dual-band flexibility—all at bargain pricing in case a model gets lost or abused!

Backcountry Hunters

Rugged water resistance and 9-hour+ battery life of the UV-82 make it easy to recommend. As does 8 watts of power to punch through heavy tree cover when positioning other members of your group.

RV Camping Club

Lots of flexibility is needed in this case - calling from campsites back to rigs near shore power, communicating as groups to tour wineries or attractions. The UV-9R Plus shines here with superior battery life, easy absolute button control (no tiny LCD menu navigation), and flexibility to program geographic channel groupings.

Disaster Response Groups

Ramp up capability even further for volunteer groups supporting disaster sites and events. The BF-F8HP checks the boxes with its ultra-clear receiver, 8 watts of transmit power, and computer customization. Screen out interference with sub-codes and make full use of those 132 programmable channel memories!

Hopefully, these real-world use case examples provide tips on identifying and matching your essential requirements to the ideal model. Avoid overbuying, but recognize where paying a small premium nets significant feature advantages over entry models.

1.4 Programming and Customization

Out of the box, Baofeng radios only access essential functions. Customized programming is necessary to unlock your use case's full potential. Two main methods are manual programming through the radio interface and advanced computer programming.

Manual Programming

The radio interface allows setting frequencies, channels, codes, and options using the menu system. This involves assigning transmit/receive frequencies to each channel and CTCSS/DCS tones or DMR IDs per channel. It works well for basic requirements but can be tedious to enter extensive data.

Manual Programming: Step-by-Step Guide

Let's quickly review the steps to program your radio manually. However, the exact buttons and menu options will vary slightly depending on your Baofeng model. Refer to your radio's manual for specifics.

Here are the steps to follow when programming your Boafeng radio manually:

1. Enter Frequency Mode:

◆ Usually, press a button labeled "VFO/MR" to switch between Channel and Frequency modes.

2. Set the Frequency:

◆ Use the keypad to enter the desired transmit frequency.

◆ Some radios require setting the receive frequency separately.

3. Set CTCSS/DCS (if needed):

◆ Press the "MENU" button, then locate the CTCSS/DCS setting (refer to your manual for the exact menu number).

◆ Use the number keys or up/down arrows to select the correct tone code.

4. Save to a Channel:

◆ Press "MENU" followed by a menu option to save the settings to a channel (the option is usually labeled "MEM" or similar).

◆ Use the number keys or arrows to choose a channel number.

◆ Press "MENU" to confirm.

5. Repeat for Other Channels:

◆ Manually program as many channels as you need.

Additional Customization

Channel Names: Some Baofeng models allow setting names for channels instead of just numbers. Look for this in the menus.

Power Levels: Adjust transmit power (high/low) within the settings.

Other Options: Depending on your model, you might have options for bandwidth, squelch levels, and more. Explore your menus!

Computer Programming

Software like CHIRP radically simplifies programming and enables advanced customization.

Key benefits include:

1. Importing Channels: Load complete channel/frequency lists from CSV files instead of manually entering them. This enables quick configuration of banks matching team/location needs.

2. Intuitive Channel Plans: Organize channels into labeled folders or groups instead of relying on vague numbers. Now, channels have intuitive names.

3. Configuration Profiles:Create reusable radio configuration templates for everyday use cases to sync across units. This is great for coordination.

4. Advanced Options: Fine-tune settings like transmit power controls, signal squelch, emergency alarms, and more to suit your needs.

Essentially, CHIRP software unlocks commercial-grade features without enterprise pricing. These include syncing specific frequency plans across locations, utilizing tones to coordinate channel sharing for large groups, scanning custom frequency ranges, and programming quick dial emergency functions.

Computer Programming: Step-by-Step Guide

Below are the steps to follow to carry out computer programming for your Baofeng radio:

1. Get the software: Download and install CHIRP from their website:

(https://chirp.danplanet.com/projects/chirp/wiki/Download).

2. Connect your radio:

- ◆ Turn off your Baofeng radio.
- ◆ Get a compatible programming cable (check your radio's model).
- ◆ Plug one end of the cable into your computer's USB port.
- ◆ Plug the other end into the programming port on your radio.

3. Launch CHIRP: Open the CHIRP software.

4. Download data (optional): If your radio is already programmed:

◆ Click "Radio" -> "Download from Radio".

◆ Select the correct port and radio model.

◆ Click "OK" and follow the on-screen instructions.

5. Edit channels:

a) Double-click on a channel to be able to edit its settings:

◆ *Frequency:* Enter the transmit and receive frequencies.

◆ *CTCSS/DCS:* Set tones if there is a need for that.

◆ *Name:* Give the channel a descriptive name.

b) You can use the menus and buttons in CHIRP to explore other settings.

6. Import channels (optional):

◆ If you have a list of channels, you can import them from a CSV file (You can locate this online or create your own).

◆ Click "File" -> "Import".

7. Upload to radio:

◆ Click "Radio" -> "Upload to Radio".

◆ Select the correct port and radio model.

◆ Click "OK" and follow the on-screen instructions.

Important Notes:

It would help if you took note of the following:

Compatibility: Make sure that CHIRP supports your specific Baofeng model.

Drivers: Your computer might require drivers for the programming cable. Check the cable manufacturer's website.

Understand that CHIRP has many advanced settings. You can explore them if you wish to customize your radio further.

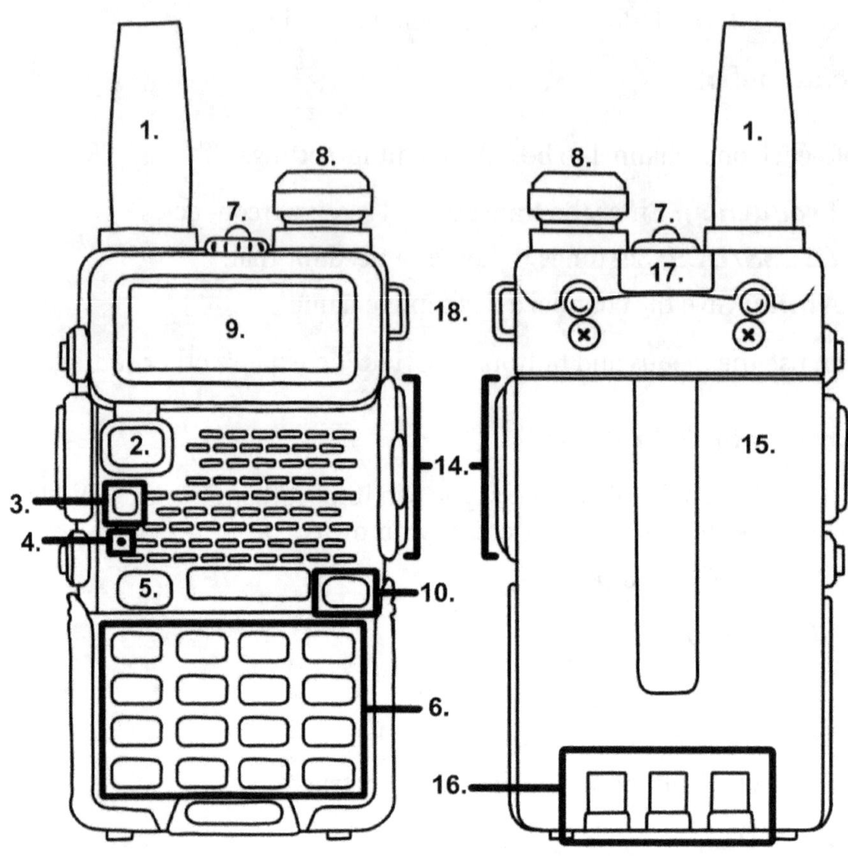

1- ANTENNA
2- FREQUENCY MODE /
   CHANNEL MODE
3- LED INDICATOR
4- SP / MIC
5- FREQUENCY DIAPLAY
   SWITCHES
6- KEYPAD
7- FLASHLIGHT
8- KNOB (ON / OF, VOLUME)
9- LCD DISPLAY

10- BAND KEY
11- SK-SIDE KEY1 / CALL
    (RADIO, ALARM)
12- PTT KEY (PUSH-TO-TALK)
13- SK-SIDE KEY2 / MONI
    (FLASHLIGHT, MONITOR)
14- ACCESSORY JACK
15- BATTERY
16- BATTERY CONTACTS
17- BATTERY REMOVE BUTTON
18- STRAP BUCKLE

BAOFENG UV-5R MODEL

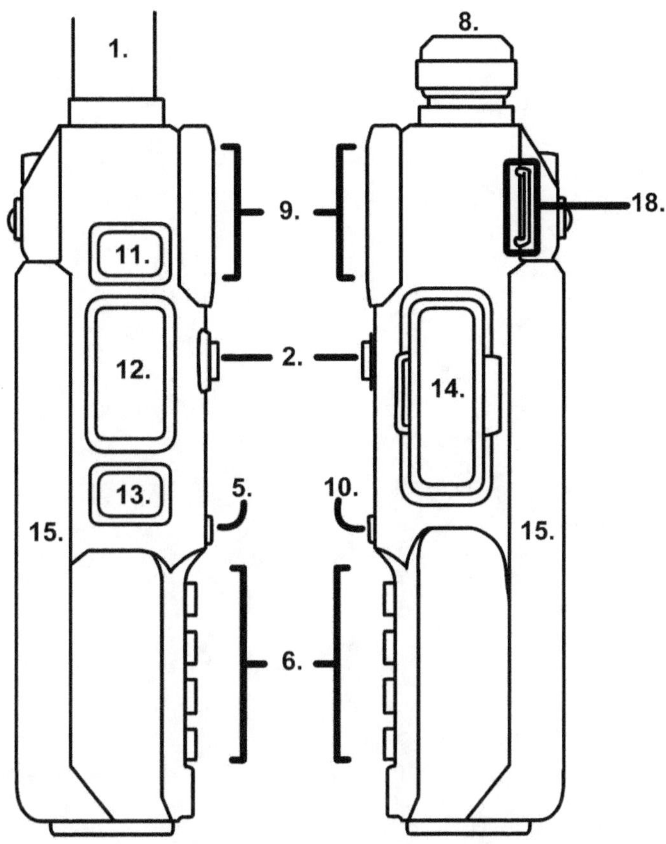

1- ANTENNA
2- FREQUENCY MODE / CHANNEL MODE
3- LED INDICATOR
4- SP / MIC
5- FREQUENCY DIAPLAY SWITCHES
6- KEYPAD
7- FLASHLIGHT
8- KNOB (ON / OF, VOLUME)
9- LCD DISPLAY
10- BAND KEY
11- SK-SIDE KEY1 / CALL (RADIO, ALARM)
12- PTT KEY (PUSH-TO-TALK)
13- SK-SIDE KEY2 / MONI (FLASHLIGHT, MONITOR)
14- ACCESSORY JACK
15- BATTERY
16- BATTERY CONTACTS
17- BATTERY REMOVE BUTTON
18- STRAP BUCKLE

BAOFENG UV-5R MODEL

CHAPTER 2: UNDERSTANDING BAOFENG FEATURES

Chapter 1 covered the basics of Baofeng radios, the most popular models, how to choose the right one and programming methods. Now, let's dive deeper into the functionality and capabilities packed into these pocket-sized devices!

I want to explore the key features step-by-step so you understand how to take advantage of the flexibility. We'll start by getting familiar with the physical buttons and display. Then, dive into clever power management tactics before entering the emergency alert capabilities.

2.1 Display, Interface, and Navigation

As noted earlier, Baofeng radios resemble basic walkie-talkies with a small monochrome display crammed between an antenna and speaker. But don't let the tiny screen fool you...it packs a ton of status indicators and menu navigation!

The display typically shows the current channel along with a 4-segment battery indicator. You'll also see icons if you have the FM radio activated, the keypad lock enabled, or the automatic power-save mode activated. It takes some memorization to get comfortable with the tiny status icons in the limited pixel space!

There are also seven physical buttons adjacent to the display that serve essential functions:

- ◆ KNOB (Power Button) – Turns unit on/off and is often used to exit menus.
- ◆ MENU BUTTON – Access configuration menus and settings.
- ◆ DIRECTIONAL BUTTONS – Moves cursor to select menu options.
- ◆ CALL BUTTON – Quick access to Channel 16 emergency/hailing frequency.
- ◆ TRANSMIT BUTTON – Holds to talk on the selected channel.
- ◆ MONITOR BUTTON – Disables squelch temporarily to hear very weak

signals.

◆ VFO/MR BUTTON – Switches between Variable Frequency Oscillator (Manual tuning) and Memory Recall (Preset channels) modes.

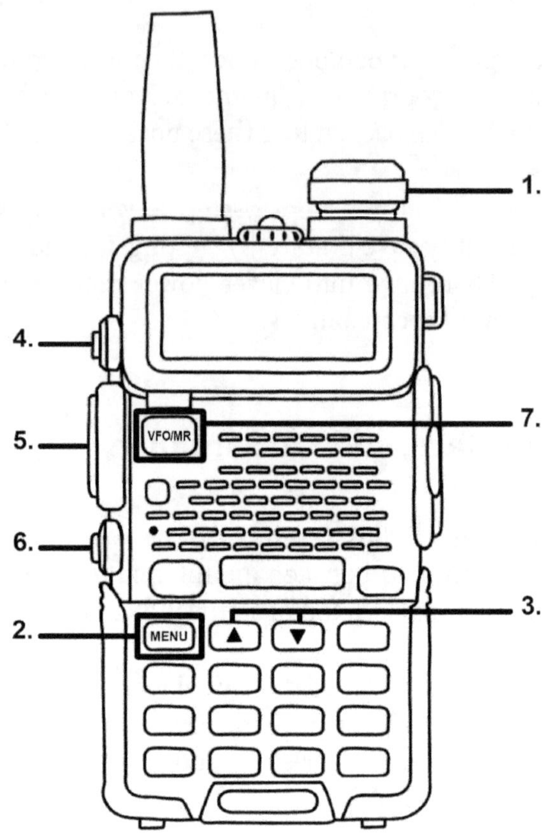

1- KNOB (ON / OFF, VOLUME)
2- MENU BUTTON
3- DIRECTIONAL BUTTONS
4- CALL BUTTON (SK-SIDE KEY1)
5- TRANSMIT BUTTON / PTT KEY (PUSH TO TALK)
6- MONITOR BUTTON (SK-SIDE KEY2)
7- VFO/MR BUTTON

BAOFENG UV-5R MODEL

Between these seven buttons and the menu system, you control every function the radio provides. It does take practice to navigate since you're scrolling through hierarchical menus on a small display using ambiguous icons.

For example, to switch the power output from high to low, you'll need to:

◆ Press Menu to enter setting mode.

◆ Scroll to the Radio Set menu.

◆ Choose TX Power sub-menu.

◆ Toggle the setting to Low.

This is why some argue that Baofeng units trade a bit of usability for flexibility. But with some familiarization, you'll soon navigate menus intuitively. And it sure beats carrying multiple single-purpose radios!

Some models, like the UV-9R, have dedicated buttons for FM radio, flashlight, and scan modes. This avoids menus for everyday actions. But most operations still utilize screen navigation and multiple button presses—part of the tradeoff for packing in maximum configurability.

The key is investing upfront. Have the user manual handy and practice all functions before entering the field. After some hands-on familiarization, menu actions will soon become second nature.

2.2 The Baofeng Radio Menu Options

The Baofeng radio offers an extensive menu of options, providing users with versatile functionalities to cater to various communication needs. With its user-friendly interface, navigating through the menu is intuitive, allowing users to access many features effortlessly. Here are the 40 different menu options available on the Baofeng radio:

No.	Menu	Function/Description	Available Settings
0	SQL - Squelch Level	Controls how sensitive the radio is to incoming signals. A higher setting filters out weak signals and background noise, while a lower setting allows fainter transmissions to be heard.	0-9
1	STEP - Frequency Step	Determines the increment by which the radio tunes up or down in frequency. Smaller steps allow for finer tuning, while more giant steps make scanning faster across a wide range of frequencies.	2.5 / 5 / 6.25 / 10 / 12.5 / 25 kHz
2	TXP - Transmit Power	Sets the radio's output power. Use HIGH for longer distances, but be aware that it drains the battery faster. LOW is better for short-range communication.	High / Low
3	SAVE - Battery Save Mode	Conserves battery power by putting the radio into a low-power state between transmissions.	OFF / 1 / 2 / 3 / 4
4	VOX - Voice Operated Transmission	Enables hands-free operation. The radio will automatically begin transmitting when it detects your voice.	OFF / 0-10

5	WN - Wideband/ Narrowband	Select the channel bandwidth. Wideband provides better audio quality, while narrowband allows more channels to fit within a given frequency range.	WIDE / NARR
6	ABR - Display Illumination Time	This controls how long the backlight remains on after a button press. Longer durations improve visibility in low light but use more power.	OFF / 1-5 secs
7	TDR - Dual Watch/Dual Reception	Enables the radio to monitor two channels simultaneously.	OFF / ON
8	BEEP - Keypad Beep	Turns the audible beep on or off when buttons are pressed.	OFF / ON
9	TOT - Transmission Time-out Timer	Limits the duration of a single transmission to prevent accidental channel blockage.	15 / 30 / 45 / 60 - 600 seconds
10	R-DCS - Receive Digital Coded Squelch	Filters incoming signals based on a digital code, allowing selective reception of specific transmissions. Note: Not all repeaters requiring a tone for access transmit a tone back to you. Leave this function turned OFF unless you are sure it is needed.	OFF / D023N - D754I

11	R-CTCS - Receive Continuous Tone-Coded Squelch	Similar to R-DCS, but uses an analog tone instead of a digital code.	OFF / 67.0 - 254.1 Hz
12	T-DCS - Transmit Digital Coded Squelch	Adds a digital code to your outgoing transmissions, allowing receivers with matching R-DCS settings to hear you.	OFF / D023N - D754I
13	T-CTCS - Transmit Continuous Tone-Coded Squelch	The analog tone equivalent of T-DCS.	OFF / 67.0 Hz - 254.1 Hz
14	VOICE - Voice Prompt	Enables spoken announcements for channel numbers or settings.	OFF / ON or OFF / ENG / CHI
15	ANI-ID - Automatic Number Identification	Transmits a caller ID-like code with your transmission. Sent when PTT is pressed and released.	
16	DTMFST - DTMF Tone of Transmitting Code	Enables the sending of DTMF tones (like on a telephone keypad) during transmission.	OFF / DT-ST / ANI-ST / DT+ANI
17	S-CODE - Signal Code	Add specific numbers to your transmission for extra identification or filtering.	1 - 15 groups

#			
18	SC-REV - Scan Resume Method	This determines how the radio behaves when it finds a signal during a scan (e.g., pauses for a set time, resumes after signal drops, etc.)	TO / CO / SE
19	PTT-ID - Push-to-Talk ID	Transmit a brief signal, potentially containing your identification code, at the start or the end of your transmission. OFF - No ID is sent. BOT - An ID is sent at the beginning of Transmission. END - An ID is sent at the End of Transmission. BOTH - An ID is sent at BOT and EOT.	OFF / BOT / EOT / BOTH
20	PTT-LT - Push-to-Talk ID Delay	Sets a delay before the PTT-ID signal is sent.	0 - 30 ms or 0 - 50 ms
21	MDF-A - Channel Mode A Display	Customizes what information is shown on the display for Channel A (e.g., frequency, channel name, etc.)	FREQ / CHAN / NAME
22	MDF-B - Channel Mode B Display	Same as MDF-A, but for Channel B.	FREQ / CHAN / NAME

23	BCL - Busy Channel Lockout	Prevents you from accidentally transmitting over an ongoing conversation.	OFF / ON
24	AUTOLK - Automatic Keypad Lock	Locks the keypad after a period of inactivity to prevent accidental button presses.	OFF / ON
25	SFT-D - Frequency Shift Direction	Used for repeater operation. Determines whether your transmit frequency is shifted higher or lower relative to the receive frequency.	OFF / + / -
26	OFFSET - Frequency Shift Amount	Specifies the amount of frequency shift to use for repeater operation.	00.000 - 69.990 Mhz in 10 kHz steps
27	MEM-CH - Store Channel	Saves a frequency and settings (like CTCSS/DCS codes) into a memory channel for easy recall.	000 - 127
28	DEL-CH - Delete Channel	Removes a saved channel from the radio's memory.	000 - 127
29	WT-LED - Standby LED Color	Let you choose the LED indicator's color when the radio is in standby mode.	OFF / BLUE / ORANGE / PURPLE
30	RX-LED - Receive LED Color	Sets the LED indicator's color when the radio receives a signal.	OFF / BLUE / ORANGE / PURPLE
31	TX-LED - Transmit LED Color	Determines the color of the LED indicator when the radio is transmitting.	OFF / BLUE / ORANGE / PURPLE
32	AL-MOD - Alarm Mode	Configures how the radio's alarm function behaves (sound type, volume, etc.).	SITE / TONE / CODE

33	BAND - Frequency Band Selection	Allows you to switch between significant radio frequency bands, such as VHF or UHF. This is similar to the [BAND] button.	VHF / UHF
34	TDR-AB - Transmit Selection while Receiving	When the dual watch is active, determine which channel (A or B) will be used for transmitting.	OFF / A / B
35	STE - Squelch Tail Elimination	Reduces the brief burst of noise heard at the end of a transmission.	OFF / ON
36	RP-STE - Squelch Tail Elimination in Repeater Mode	The same function as STE but is specifically tailored for repeater use.	OFF / 1, 2, 3 - 10
37	RPT-RL - Delay the Signal Receiving to Repeater	Introduces a slight delay before retransmitting through a repeater, which can help prevent interference.	OFF / 1, 2, 3 - 10
38	PONMSG - Power-on Message	Customizes the text or image briefly displayed on the screen when the radio powers on.	FULL / MGS
39	ROGER - Roger Beep	It turns on or off a brief "beep" sound at the end of your transmission to signal that you've finished talking.	OFF / ON
40	RESET - Restore to Factory Default Settings	VFO: resets all menus to factory defaults. ALL: reset the radio's settings and clear any stored channels to their original factory defaults.	VFO / ALL

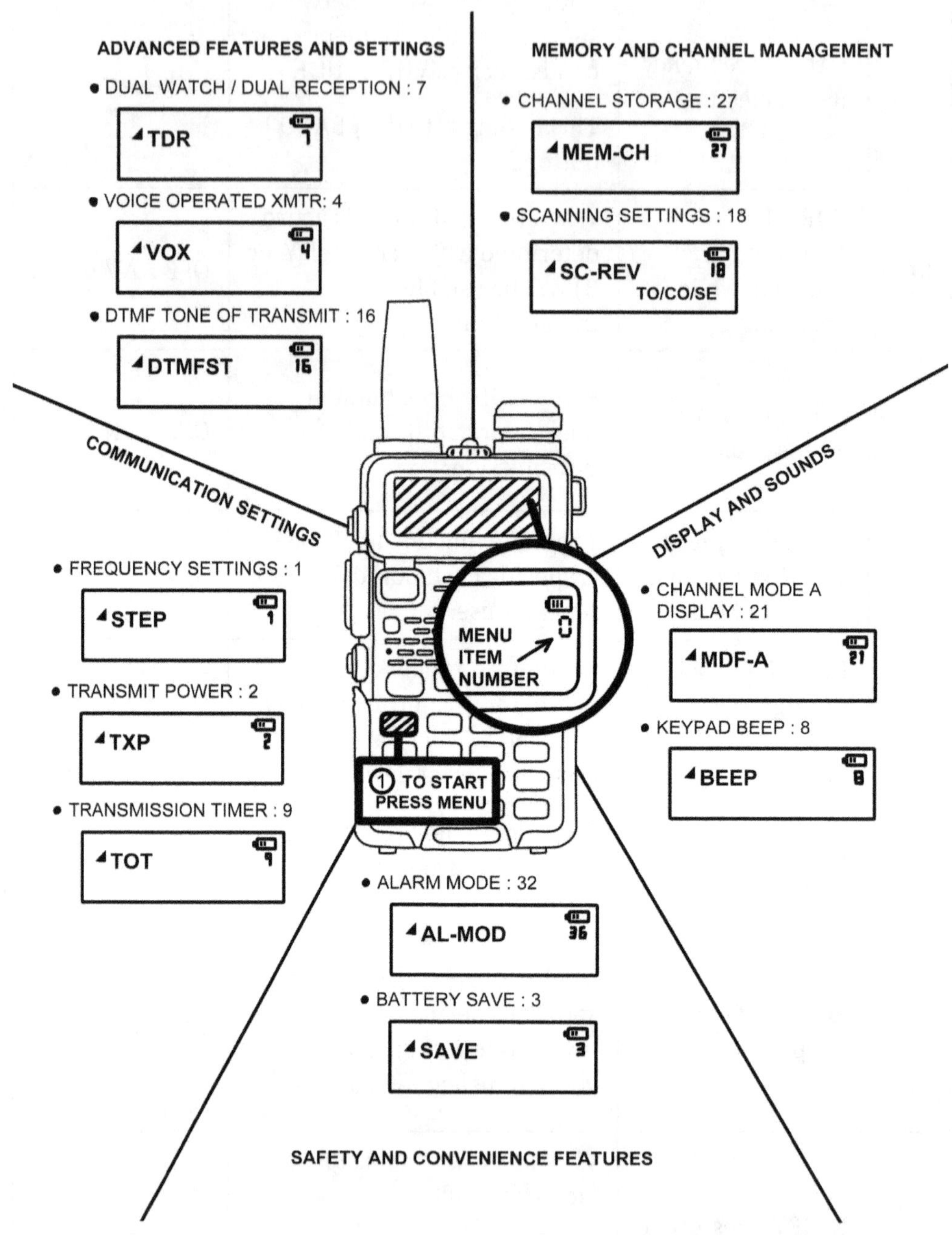

MENU OPTIONS FLOWCHART

2.3 Power Management and Battery

Being smart with power management pays big dividends for portability and convenience when relying on any battery-powered devices in the field. Baofeng packs a few helpful features to optimize consumption so you can communicate longer.

First, models like the UV-5R support several battery types, like Ni-MH, alkaline, or lithium-ion packs. Just ensure the polarity is aligned and enjoy hours of operation. For extended trips, always pack spare charged batteries!

Baofeng batteries are measured in milliamp hours (mAh). The standard capacity is around 1800 mAh, but higher-density 18650 packs up to 5000 mAh are available. The higher the number, the longer the potential operating time.

Whether to pack more small batteries or a couple of high-capacity packs on longer excursions is constantly debated. Supplementing a big 5000 mAh battery with some lightweight 1800 mAh spares provides a nice balance for flexibility. Quickly swap the little ones when convenient instead of stressing about runtime.

Even with a thrifty Chinese radio, the classic advice still applies—transit quickly, if possible, between receiving and transmitting. Constant broadcasting chews through charge rapidly. But judicious transmission supplemented by ample reception optimizes the battery.

Let the burst tones do their job for channel clarity, too. Resist holding down transmit buttons excessively long to send Morse code-style spaced bursts. The radio squelch handles quick relay handoffs between partners just fine!

Take advantage of the built-in battery-saver function to extend passive operating time, too. This mode puts the radio in low-power sleep until detecting channel activity. Just a few tweaks extend that 1800 mAh pack considerably!

Finally, top off every chance you get, even if the battery shows half full. For example, I toss my UV-5R on a charger in the truck while driving between sites, plug it into a backup USB battery in camp, and start each day fully charged without worrying about aggressive use destroying runtime.

Pair battery discipline with sufficient spares for your expected usage. That peace of mind means focusing on tasks instead of being distracted by dying batteries!

2.4 Emergency Features

Emergency capabilities are where Baofeng models shine over most commercial-grade walkie-talkies. They seamlessly interoperate with many ham and GMRS frequencies vital for coordinating urgent needs between companions or responding to volunteer groups.

All units include a dedicated call button providing instant access to Channel 16 - the international hailing and distress frequency. Press to instantly call nearby stations to relay emergency messages.

This is the equivalent of maritime VHF channel 16, the contact channel for initial calls to other stations. Any station hearing a distress call on channel 16 will shift to working channel 9 to coordinate assistance and relief.

Baofengs also supports NOAA weather radio broadcasts. This integration is invaluable for receiving location-based weather alerts, allowing you to take cover or warn the group before storms strike unexpectedly. Set your SAME code to match state and county to enable targeted local alerts.

And we must remember the all-important flashlight built into most units! This may sound mundane, but it is invaluable for illuminating surroundings when coordinating nighttime emergency responses or recovering gear after dark.

Custom emergency signaling capabilities are also possible through creative programming...

Establish a designated emergency channel with matching sub-codes so you can access your party directly without broadcasting openly. Transmit a quick preset distress Morse code burst, distinguishable from standard chatter by the unique pattern.

Set another channel to constantly scan emergency and weather frequencies so you catch coordination broadcasts between responders after initial check-ins. Those are excellent capabilities!

Get creative by programming a unique alarm tone from the speaker, even when muted. Now, group members can instantly signal an emergency blast to other radios in their pockets well before transmitting!

Too many amateur operators need to pay more attention to emergency planning with their radios. But Baofeng's flexibility shines here by integrating critical features like NOAA tuning, hailing frequencies, and programmable alerts. Use these tools to remain situationally aware, signal others, and request assistance when dire situations arise far from help.

That concludes the extended tour of all the ins and outs of standard Baofeng radio capabilities. Hopefully, you feel more familiar with the options after this overview. The tiny interfaces hide tons of flexible functionality!

Let's immediately program your specific model with emergency channels, signaling options, and other handy functions.

CHAPTER 3: SETUP, ACCESSORIES, AND FIRST USE

Let's shift gears to discuss the practicalities of getting your Baofeng radio up and running! I'll cover everything from initial test procedures to charging batteries, assembling antennas, and that all-important first power on.

No matter how much background knowledge you gain about capabilities, nothing substitutes for hands-on experience. So, let's walk through best practices step-by-step to get familiarized and avoid any rookie mistakes on activation day!

3.1 Initial Setup Steps

I recommend finding a table at home with ample elbow room for unpacking before an expedition. There is no sense sitting on the tailgate reading manuals as storm clouds roll in!

Let's walk through unboxing your new Baofeng Radio.

When you the box, you'll find several pieces that need to be put together before using the radio. Here's an overview of what's inside and how to set things up:

What's in the Box

- ◆ The radio itself - this central handheld unit is what you'll hold and talk into.
- ◆ A battery pack - provides portable power for the radio.
- ◆ A belt clip - attaches to the back of the radio so you can clip it to your belt.
- ◆ A charging base - is used to recharge the battery pack.
- ◆ An antenna - lets you transmit and receive radio signals.
- ◆ An earpiece - can plug into the radio to hear audio privately.
- ◆ A user manual - teaches you how to use all the features.

Here is the quick process:

1.   Carefully remove the radio, battery, charger, and components from the box to avoid damage.

2.   Visually inspect for any cracks or defects; test buttons and knobs function.

3.   Attach the antenna to the connector mounted on top until finger-tight.

4.   Review manuals and accessory guides to understand hardware.

5.   Fully charge the battery for at least 5 hours before first use.

6.   Power on to confirm the display activates and test the speaker.

7.   Sync the clock/calendar and begin manual programming for the activity.

8.   Scan the user manual and practice every function to build familiarity.

Let's break the process in a more detailed form for smooth sailing:

Attaching the Antenna

You'll notice a small hole at the radio's top labeled "Antenna." Take the included antenna and line it up with the hole. Gently twist the base clockwise until it doesn't turn any further - that means it's securely installed. Don't twist it by the top section, or you could break the antenna.

ANTENNA INSTALLATION

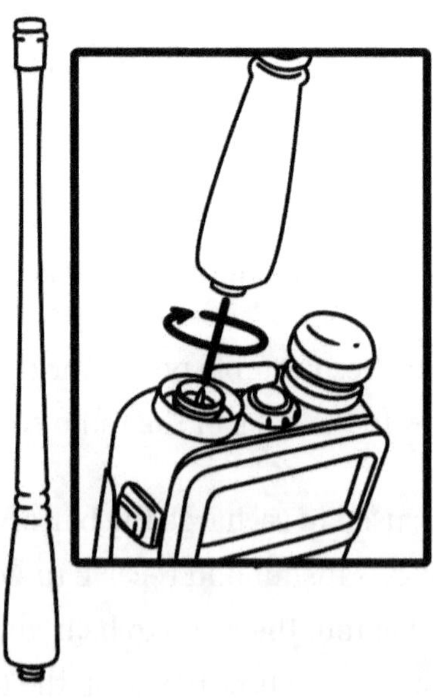

Installing the Battery

Batteries supply the operating power for portable radios. Many Baofeng models use a removable battery pack. Slide the battery into the open back of the radio, making sure to line up the metal contacts. You'll feel it click into place when properly seated.

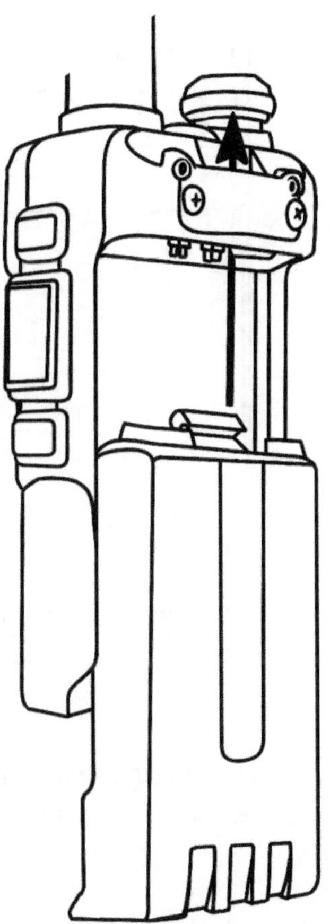

Attaching the Belt Clip

The back of your radio has a spot to screw on an included belt clip, allowing you to clip the unit securely onto a backpack strap, pocket edge, or your pants belt. This makes carrying the radio super convenient while keeping your hands free!

BELT CLIP FASTENING

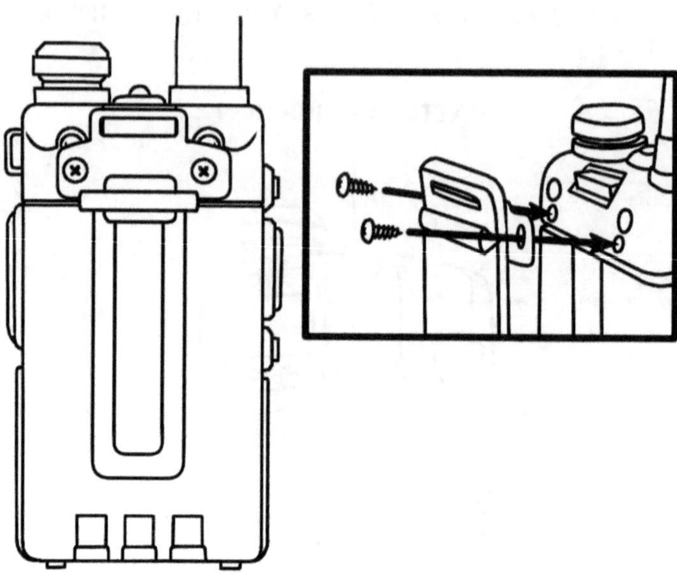

Installing the Headset

The Baofeng radio box also contains a wired headset with a microphone. This lets you listen to incoming audio and speak into the mic privately without holding the radio.

To install it, locate the connector jack along the top edge of the radio body. It will match the plug found on the headset cable. Gently push the headset plug into the radio jack until it clicks securely.

You can now put the earbuds in and adjust the mic boom arm so the microphone sits about an inch from your mouth. The external headset becomes the primary speaker and mic for exchanging radio communications. The radio's built-in mic and speaker automatically get overridden.

Don't jump straight to advanced computer programming or field trials. Mastering the equipment basics through experimentation instills familiarity and confidence in communicating later when emotions run high!

Secure a compatible second radio and extra battery or program cable if possible.

Quick access to spares and the ability to clone settings across units will save you tremendous time and effort.

HEADSET WIRING

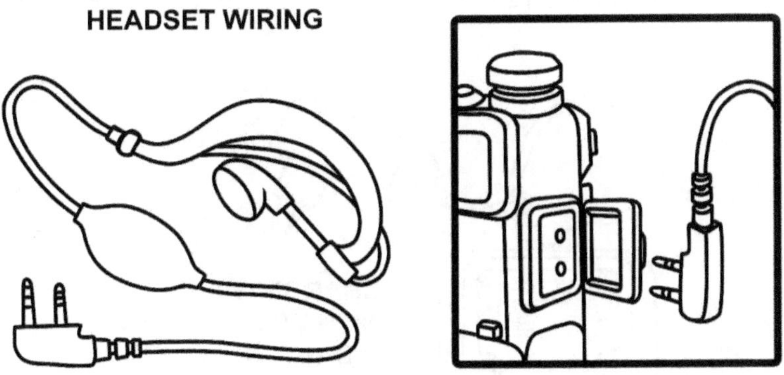

3.2 Batteries and Charging

Baofeng batteries are measured in milliamp hours (mAh), indicating potential operating time. Standard batteries ship around 1800 mAh, but upgraded high-capacity packs reach 5000 mAh for longer runtimes.

Before using a new Baofeng radio, you'll want to give the included battery a full 5-hour initial charge. That ensures maximum battery capacity right off the bat! Use the included drop-in charging base by connecting it to a wall outlet using the power adapter. Place the radio with the battery installed in the cradle—the LED lights will indicate the charging status so you know when it's complete.

Confirm your charger outputs the correct voltage matching the battery type before connecting each time. Misaligned charging attempts risk permanent damage! Check markings molded into packs carefully.

I prefer trickle-charging batteries overnight to quick top-off cycles to maximize battery life. But balancing charge needs on longer excursions requires packing multiple interchangeable spares. Just remember polarity orientation when swapping!

Also, consider supplemental USB battery packs when convenient outlet access allows. For example, stashing a 20,000 mAh USB supply in base camps keeps radios perpetually humming without draining precious onboard cells.

Finally, like your smartphone, take precautions to avoid over-discharge. Allowing Li-ion packs to dip too low increases the risk of premature failure. Know the warning signs of low power, like dim displays and cutoff failures. Keep a few 1800 mAh cells as emergency spares for gracefully powering down before total shutdown.

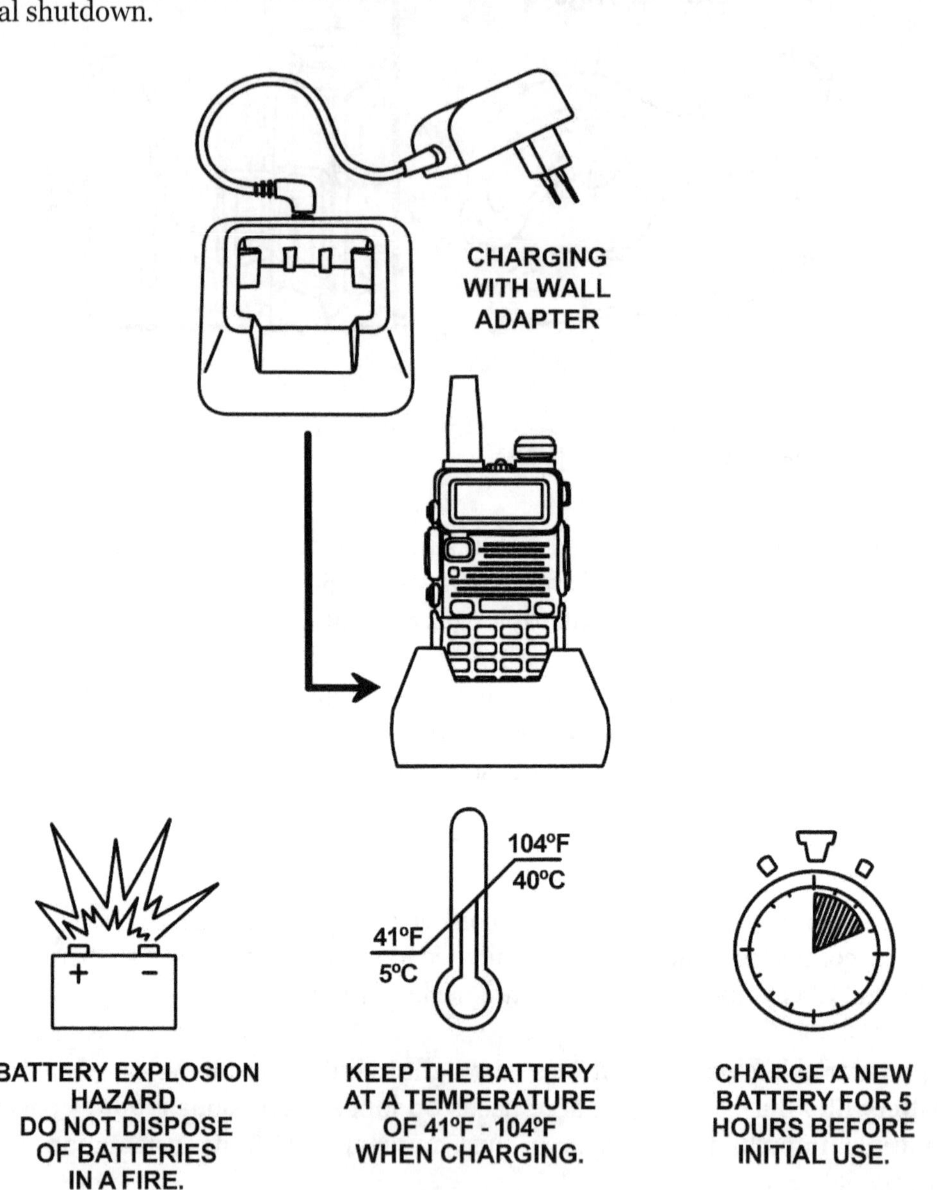

KEEP YOUR BATTERY HEALTHY

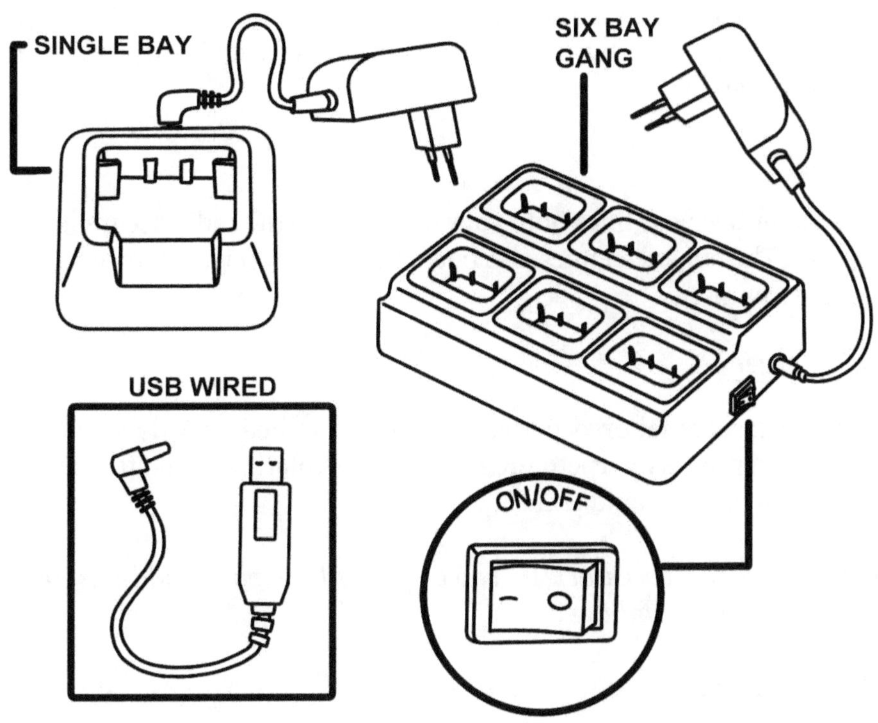

VARIOUS CHARGER STANDS

3.3 Essential Accessories

Beyond just extra cells, some essential accessories radically enhance real-world functionality:

Spare Antenna

Your primary antenna is essential so losing it will make your Baofeng model useless, irrespective of the model. Always pack multiple antennas, especially if using temporary magnetic mounts. A small wire signal stick stashed in the go-bag saves the day!

Headset

Scanning emergency frequencies via speaker proves impractical for long-term monitoring. A headset with an in-line microphone frees up hands while allowing borderline readability.

Speaker Mic

Clipping larger over-ear mics onto your collar keeps the radio secure in a pouch while active. The enhanced mics also boost transmit/receive clarity considerably in noisy environments!

Desktop Charger

Multi-bank chargers tailored perfectly for these radios provide safe overnight charging to keep packs perpetually fresh. When rotating off-shift, leave one in the hotel room or command tent.

Beyond standard accessories, some special-purpose add-ons exist, unlocking unique capabilities:

Throat Microphones

Perfect for high-noise environments, rendering headsets ineffective. Throat mic vibrations transmit speech directly, avoiding ambient interference. Allows clear communication in loud machinery spaces. Downsides include difficult placement and training for effective positioning.

Surveillance Earpieces

Extremely tiny in-ear receivers plus flush-mount boom mics provide ultra-discrete radio use, avoiding attention. They are used extensively by protection officers and government agents, maintaining low profiles in public. Despite the visual effectiveness of visually masking radioactivity, they can be challenging to insert/secure comfortably.

VOX Voice Activated Headsets

Headsets with built-in VOX circuitry toggle transmit modes automatically by detecting speech patterns instead of using manual buttons. This allows hands-free radio communication activated on voice alone. Drawbacks include increased battery drain potential and background noise keying unwanted transmissions. Proper gain settings mitigate over-triggering but reduce effectiveness in very loud environments.

High Gain Whip Antennas

Whip or telescoping external antennas focus signal strength, extending radio reach considerably compared with built-in antennas. But at the cost of antenna fragility, risking damage, unwieldy length catching on surroundings, and difficulty masking such prominent communication accessories visually. Require durability versus performance evaluation assessing projected use conditions.

While these specialty accessories fill valuable niches, tailoring radios to unique applications ensures you first master standard-equipped hardware and usage fundamentals covered previously. Only after core radio competency is achieved should you consider augmenting functionality with these specialty accessories. Their powerful capabilities enable advanced applications, but unnecessary complexity risks hampering fundamentals for most users.

Hopefully, these practical setup insights will help you smoothly kick off your exciting journey into Baofeng radios. We covered everything from hardware familiarization to battery charging best practices and critical accessories to have on hand.

You're equipped with enough background to begin experimenting first-hand with all those neat features, from squelch tweaks to emergency signaling.

CHAPTER 4: BASIC RADIO COMMUNICATION

Now that we've prepped and prepared your new Baofeng radio and accessories let's focus on fundamental communication.

I'll walk through powering up, adjusting squelch, selecting channels, pressing buttons to talk and listen, dialing in volume, clarifying microphone technique, and more—everything needed for crisp exchanges with your partners near and far.

This chapter aims to smooth the first transmissions while building vital competence in practicing radio protocols. Master this first before we customize programming and exploit more advanced functions!

4.1 Turning On & Basic Radio Controls

Let your Baofeng adventure begin! Power up by pressing and holding the red button at the upper left. You see the radio model splash screen and a channel/ frequency display.

Next, rotate the knob left of the screen counter-clockwise to turn the volume up to 5 or so. Now, press the monitor button (lightning icon), and speak clearly to the top-front microphone. Smile as your voice transmits through the speaker!

Playing with basic radio controls first helps get familiar before communications begin:
- ◆ Volume Knob - Rotating left/right adjusts receive volume.
- ◆ Squelch Knob - Increases/decreases microphone sensitivity.
- ◆ Power Button - Press for on/off and to exit menus.
- ◆ Transmit Button - Hold this side button to talk.
- ◆ Monitor Button - Temporarily check background noise.
- ◆ Band Switch - Toggles between VHF and UHF bands.
- ◆ Antenna - Transmits/receives signals; keep clear.
- ◆ Speaker - Outputs received voices or tones.

Don't stress memorizing now as we break down individual capabilities more soon. Just get a feel for how Buttons and knobs influence operation through hands-on practice.

Basic Radio Operation Using the UV-5R

Here are basic radio operations using the UV-5R model. These operations are almost the same as those for every other model.

Turning the Radio On and Off

To turn the radio ON: Turn the round button on the top clockwise until it clicks.

You'll see the screen light up - it's now powered on!

To turn it OFF: Turn that same top button counter-clockwise until it clicks again. The screen will go black, indicating it's safely powered off.

Adjusting the Volume

Increase: To increase the speaker volume, turn the volume knob clockwise. Keep turning it to make the audio playback even louder.

Decrease: Turn the volume knob counter-clockwise if you want to lower the volume played through the radio speaker. Keep turning counter-clockwise to make it more and more quiet.

Changing Channels and Frequencies

This radio operates on different radio frequency bands divided into numbered channels.

To choose a saved channel, Press the "VFO/MR" button to enter Channel mode. Use the number keys to enter a channel number like "15". Or use the up/down arrow buttons to scroll through saved channel numbers.

Manually Frequency Tuning

If you want to tune the frequency manually, press "VFO/MR" to enter Frequency mode. Use the number keys to enter a frequency like "462.7000 directly." Or use the arrow buttons to tune the frequency up/down slowly.

You can swap between the current channel/frequency using the "A/B" button. The "selector arrow" indicates if A or B is currently selected.

The "BAND" button switches between VHF and UHF frequency bands.

Locking and Unlocking Buttons

To lock the buttons (except volume and talk), hold the "#" key until you see the lock icon onscreen.

Press and hold the "#" key to unlock the buttons again. The lock icon will disappear when unlocked.

4.2 Channels and Squelch Settings

Frequencies and channels sound technical right now. Let's demystify a bit!

Frequency - Where in the radio spectrum your signals travel.

Channel - Saved frequency pairing for quick recall.

For example, transmitting on 400 MHz UHF reaches partners nearby pretty reliably. Receiving their reply on 405 MHz completes the two-way exchange.

Rather than manually dialing these transmission/reception frequencies, we assign the pair a Channel name. Now, choose Channel 2 on the radio display to access both frequencies automatically!

The radio squelch function also helps ensure communication clarity. This suppresses noise, allowing reception only when signals exceed a preset threshold. Adjust sensitivity high enough to receive partners clearly without getting drowned in static.

Dial down the volume and test rotating the squelch knob clockwise from fully counter-clockwise. At a certain point, cut-off noises as weaker signals get filtered away by the higher squelch. Leave a slight buffer so weaker but still intelligible signals through.

Here is the exact step to follow in setting the squelch:

To adjust the squelch, Press Menu -> 0 -> Menu again.

Then, use up/down arrows to increase/decrease the setting. Go up one notch at a time until the static sound completely cuts off.

4.3 Making and Receiving Calls

All right, ready for your first exchanges? Here is the proper technique:

1. Select desired channel - Use dial, menu, or dedicated button.
2. Listen briefly - Confirm the channel is free for use.
3. Press & hold transmit - Usually side orange button.
4. Speak clearly into the microphone - ~1 inch away works best.
5. Release transmit when finished.
6. Await reply on the same channel after a short pause - Provides time to switch from transmit to receive.

Simple enough in theory! But many accidental mistakes plague beginners...

The notorious "stuck mic": Occurs when you hold the transmit for too long, locking the channel. Whoops! If this happens, refresh by toggling channels or power cycling as needed.

The too-eager listener: Starting transmission before the partner fully finishes speaking. Slow down that trigger finger! Wait for a clear channel marker.

The mumbler: Comprises chewing on the mic, long pauses, and a muffled environment. Keep your lips close and project your voice! Add filler words like "Over" at the end to signal the transmitted portion.

And don't forget the volume-cocked confidence crusher: Blasted eardrums when someone replies with far hotter signal strength or closer mic than you! Keep volume conservative until actual field use establishes typical reception clarity.

Beyond these tips, basic radio etiquette goes a long way toward maximizing potential partners while minimizing annoyance. Use common sense—identify call signs, avoid long-windedness, signal clearly when finished, confirm messages received, and keep patience if others are less experienced.

The practice also perfectly copies signals through real interfering scenarios—static, bad reception, weak batteries, and loud backgrounds. Early training in noise helps the brain automatically fill in garbled messages later. It's much easier to improve now rather than when coordinating critical tasks!

Are you feeling comfortable with the foundations? Now, let's unlock serious functionality, customize channel banks, and develop advanced memory channel functions, which will be discussed in later chapters!

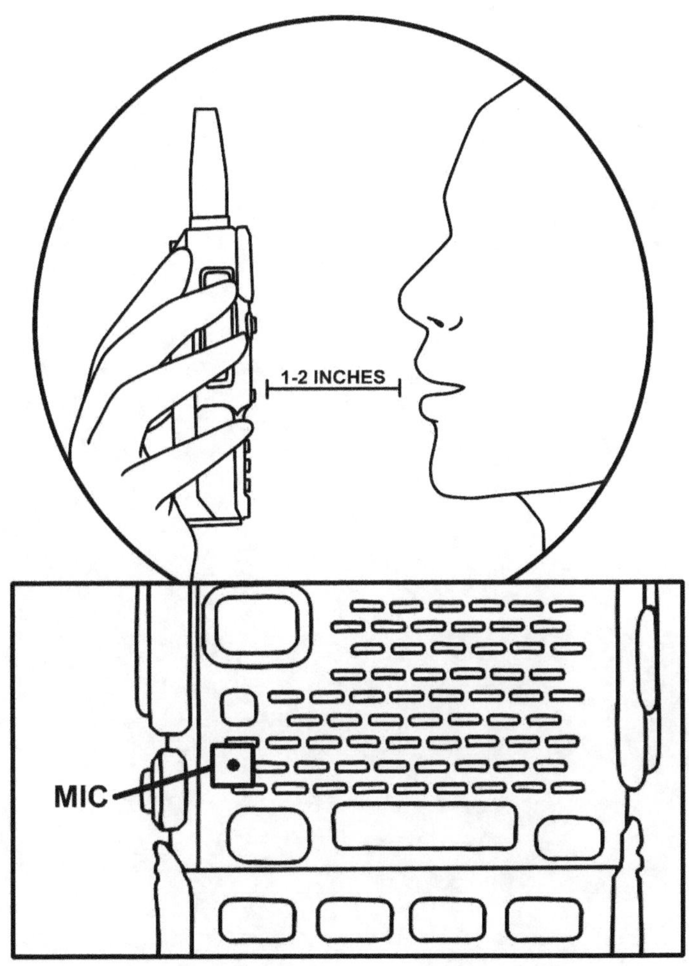

HOLDING THE RADIO CORRECTLY

CHAPTER 5: FREQUENCIES, CHANNELS AND PROGRAMMING

Let's examine radio frequencies, channel configuration, and unlocking advanced customization through programming! Since you're familiar with the Baofeng model basics, let's explore flexible radio capabilities that match your unique needs.

This chapter will discuss terms like bands, duplex vs. simplex, and software tools that simplify complex frequency management for smooth sailing. Follow along and practice key concepts to master them confidently by the end of the chapter!

5.1 Radio Frequency Bands Overview

Radio signals utilize electromagnetic waves transmitted at specific frequencies to carry information. These varying frequencies are categorized into bands suitable for particular applications based on propagation behaviors.

Your dual-band Baofeng taps into both key blocks:

VHF High Band

Typically covering 136MHz to 174MHz, the VHF High band boasts signals that can traverse longer distances with line-of-sight propagation. This makes VHF frequencies ideal for scattered user groups spanning wilderness.

UHF Band

Usually spanning 400MHz to 520MHz, the UHF band offers shorter-range signal coverage that better serves more localized users. A benefit is that UHF frequencies can better penetrate physical structures, vital for indoor applications.

Understanding the frequency ranges used by various entities allows intelligent channel planning. You should strive to group similar organizations or functions in dedicated channel segments. This minimizes confusion from too diverse a mixture while maximizing opportunities for intentional interoperability when required.

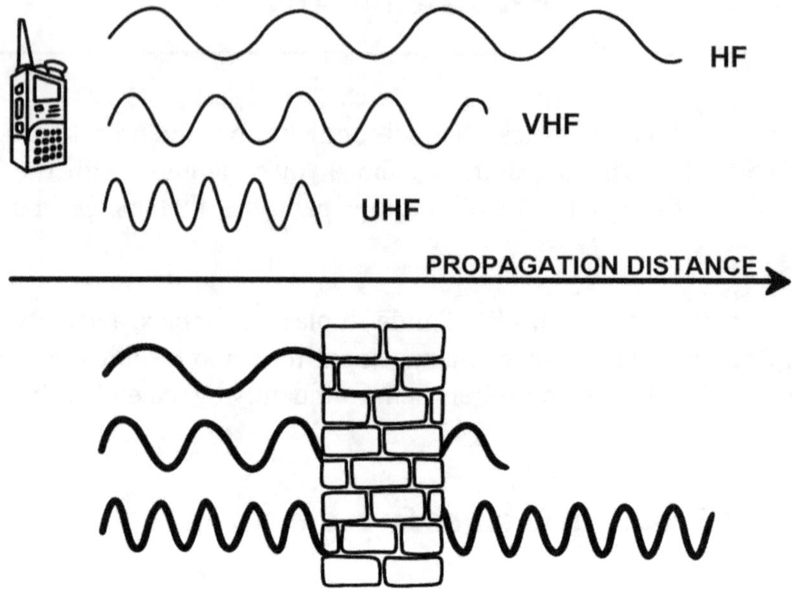

RADIO WAVE VARIETIES BY DISTANCE

Now, let's discuss "duplex" vs "simplex" modes. You may be wondering what they are. Let's cover the basics!

Duplex: Duplex means radios can transmit outbound signals on one designated frequency while simultaneously receiving inbound signals on a different paired frequency – facilitating continuous back-and-forth exchanges similar to a phone call.

Simplex: Simplex means all stations utilize a single shared frequency for alternately transmitting brief messages and then explicitly announcing "Go ahead," indicating the channel is cleared for responses. This relay-style communication demands more disciplined timing but is needed when radio hardware lacks distinct send/receive frequencies.

Let's explore an example:

Duplex Exchange: You transmit instructions to field techs on 462.5625MHz as they provide acknowledgments nearly simultaneously at 467.5625MHz. There is no need to clunkily interject "Over!" mid-statement before releasing the mic to avoid talking over replies.

Simplex Exchange: Team members must adjust the turns before sending complete directions or location reports on the share433.7000MHz before saying "Go ahead" to signal all information conveyed. This prompts the next station to key up continuing conversational bursts. It has a more controlled back-and-forth cadence but is less efficient.

Quick Frequency Tuning on the Baofeng UV-5R

If you want to tune your Baofeng and start communicating out of the box, keep it in Frequency instead of Channel Mode. Let's use the UV-5R model to illustrate a quick, simple process.

Press the "VFO/MR" button - if voice prompts are enabled, it will state either "Channel Mode" or "Frequency Mode." Having it in Frequency Mode allows the precise frequency desired to enter directly.

Frequencies require inputting six total digits, including decimals, to take effect. For example, to access the 143.100 MHz frequency, press 1-4-3-1-0-0. Entering fewer digits won't change from the current frequency.

If you incorrectly enter a frequency, simply wait a few seconds without pressing additional keys to allow the display to revert to the previous working frequency. There is no need to stress input mistakes!

Once correctly keyed in, you'll see the new 6-digit frequency reflected on the display. Use the arrow buttons to toggle between adjusting the top Primary frequency and the bottom Secondary frequency.

You can adjust the small arrow icon points at the selected frequency value. Press the "A/B" button to change the frequency at which the arrow icon points for modification between the two options.

To enter a saved Channel in Channel Mode, Press the "VFO/MR" button until it says "Channel Mode." Then, input the channel number to select the saved frequency pair for that channel.

For example, to choose Channel 12, press 1, 2. Or for Channel 2, press: 0, 2. The channel frequency selected will be displayed.

It's pretty straightforward, right? Yet first-timers often overcomplicate or get frustrated when changes don't stick. Avoid overthinking. The interface is simple if you enter 6-digit frequencies decisively and watch for the arrow icon to know which value gets altered when tapping or scrolling.

Now that we have vital concepts down, let's unlock your gear's true potential by customizing channel configurations to match your needs precisely!

5.2 Channel Programming

Channel programming refers to assigning frequency pairs to numbered memory slots in the radio labeled as "channels." Through programming, desired transmit and receive frequencies get matched to specific channel numbers for later quick recall. This means that the Baofeng allows saving specific frequencies you often use as preset memory channels, which avoids manually re-entering them repeatedly.

And since we have broken down the core terms in previous sections, programming your desired transmit and receive frequency pairs as preset channel banks tailored to your usage needs becomes easy. Here is a quick, simple step on how to save frequencies as memory channels:

1. Press "VFO/MR" entering Frequency Mode.
2. Enter your desired frequency (e.g. 400.5000 MHz).
3. Enter paired receive frequency or leave blank for simplex operation.
4. Press the "Menu" button, which shows configuration options.
5. Press "2" then "7" on the keypad to open Memory Channel Menu Item number 27 (this menu allows frequency saving).
6. Press "Menu" again, opening the list of channel slots.
7. Use arrows to select an empty channel slot without a "CH-" prefix.
8. With the empty channel selected, press "MENU" again to save your frequency to that channel.

For example, you might save 400.5000 MHz to Channel 16. "Channel 16" can be quickly reaccessed in Channel Mode without repeatedly re-entering the exact 400.5000 MHz digits.

To Verify Successful Save:

1. Press "VFO/MR" to return to Channel Mode.
2. Use arrow keys to scroll through channels until finding the channel number you saved.
3. Confirm it shows the correct frequency in MHz that you originally entered in the steps above.

Now, you can quickly access commonly used frequencies as dedicated channel shortcuts!

Manually inputting each proves tedious but quickly establishes organized banks.

Easier still leverages computer software, allowing the rapid import of shared frequency allocation lists to program entire stacks. This prevents painful manual input prone to typos or omitted decimals.

Once channel upload completes, you can intuitively browse by channel names and folder-style groupings instead of vague numbers:

Marketing > Channel 1-20.

Service > Channel 21-40.

Security > Channel 41-60.

Facilities > Channel 61-80.

Executives > Channel 81-100.

Custom sorting with descriptive identifications makes usage far more transparent for all communicating parties. Tap a name to access that channel's programmed settings rather than remember obscure number designations.

Streamlined Radio Programming for Enhanced Usability

As you might have seen above, advanced software programming unlocks exponential gains in harnessing your gear's true capabilities. While Baofeng two-way radios offer stellar flexibility, their extensive parameter adjustments often prove daunting and lack intuition. Fortunately, several software tools reduce complexity dramatically for smoother out-of-box operation.

Let's quickly examine efficient computer-assisted options to replace tedious manual programming tasks with reliable simplicity. Let's transform radio management frustration into satisfying productivity.

CHIRP Software Simplification

CHIRP programming software enables the intuitive organization of channels using descriptive names instead of numbered indices. Custom subgroups by family, event, or purpose make much more sense than recalling vague identifiers like Channel 25. Additional benefits like color coding, reordering, copying, and configuration profiles enhance day-to-day usage while eliminating repetitive setups.

USB Cable Connectivity

A USB cable links the Baofeng device to home computers running CHIRP by establishing a data transfer bridge. Once channel plans are finalized on screen, sync newly updated profiles to radios in seconds without manual keypad entry. Streamlined configurations keep groups communicating quickly.

Online Frequency Databases

Reference sites like RepeaterBook contain searchable transmission frequency listings for public agencies in most regions. Rather than monitoring exchanges to log critical data manually, look at town police and EMS channels, export listings pre-formatted for CHIRP, and sync to radios. Monitoring local first responders becomes simple when databases assist.

Spreadsheet Sharing & Syncing

CSV and Excel files allow the seamless sharing of organized frequency plans across groups. Centralize master allocated channel resources on Google Sheets so changes sync instantly organization-wide. Bulk editing on computers generates effortless updates radio-side through routine cable transfers.

So, linking cross-platform software with available public frequency data sources hugely reduces the time spent attempting manual programming. Focus more on productive exchanges.

5.3 Advanced Memory Channel Functions

Efficient entities greatly benefit from advanced channel feature configuration including defining dedicated emergency channels, activating channel scanning capabilities, and adjusting transmit power for battery conservation.

Let's explore a few examples where tailored programming pays dividends:

Scanning Enables Automatic Cycling

Instead of manually checking Activity across 20 channels, enable scan mode to automatically cycle through all programmed listings while pausing whenever traffic is detected. Fantastic for coordinated responses!

Let's quickly look at some more details on scanning your Baofeng.

Enabling the Scanner

To enable scanning, press and hold the * key for about two seconds. The radio will start cycling through the programmed channels to check for activity.

Scanning Modes

There are 3 main scanning modes to choose from:

1. **Time Operation Mode:** Pauses on the channel for 5 seconds after traffic is done before resuming scan.
2. **Carrier Operation Mode:** Resumes scan immediately after active transmission finishes.
3. **Search Operation Mode:** Must manually press * to resume scanning after it is halted.

Configuring Scanning Mode

Navigate to Menu Option 18 using the arrow keys to open the scanning setup. Then, choose the desired mode:

1. Press Menu.
2. Input 1-8 on the keypad to open the scanner mode menu item.
3. Press Menu again to enter.
4. Use arrows to select Time/Carrier/Search mode.
5. Press Menu again to save the choice.

Emergency Alerts Can Trigger Channel Checks

When activated by a matching emergency CTCSS sub-code tone, configuring a priority channel to begin scanning additional channel banks instantly improves alerts effectively. Now, your teams coordinate responses on the fly when emergency buttons prompt mobilization. That means that the radio can scan through all tone frequencies to identify CTCSS or DCS codes in use on an active channel:

1. Press Menu.
2. Input 10 for DCS tones or 11 for CTCSS tones keypad.
3. Press Menu to initiate tone scan.
4. button starts blinking as the scan engages.
5. Stops on identified tone, beep when match found.

Per-Channel Transmit Power Adjustments

Conserve battery lifespan by lowering transmit power to 1 watt for ordinary chatter across short distances. But keep emergency channel transmit power boosted to 5 watts for farther reaching urgent alerts. Why drain power shouting typical daily updates? Customize!

Pre-Programed GMRS Frequencies List

The table shows the channels that are for use with the General Mobile Radio Service (GMRS). These are channels 1-7 (5 Watts), 8-14 (0.5 Watts), 15-22 (5 Watts). While the channels 23-30 (5 Watts) are for the repeater.

Channel No.	Name	Cha. Freq (MHz)	Power
1	GMRS-1	462.5625	5 Watts
2	GMRS-2	462.5875	5 Watts
3	GMRS-3	462.6125	5 Watts
4	GMRS-4	462.6375	5 Watts
5	GMRS-5	462.6625	5 Watts
6	GMRS-6	462.6875	5 Watts

7	GMRS-7	462.7125	5 Watts
8	GMRS-8	467.5625	0.5 Watts
9	GMRS-9	467.5875	0.5 Watts
10	GMRS-10	467.6125	0.5 Watts
11	GMRS-11	467.6375	0.5 Watts
12	GMRS-12	467.6625	0.5 Watts
13	GMRS-13	467.6875	0.5 Watts
14	GMRS-14	467.7125	0.5 Watts
15	GMRS-15	462.55	5 Watts
16	GMRS-16	462.575	5 Watts
17	GMRS-17	462.6	5 Watts
18	GMRS-18	462.625	5 Watts
19	GMRS-19	462.65	5 Watts
20	GMRS-20	462.675	5 Watts
21	GMRS-21	462.7	5 Watts
22	GMRS-22	462.725	5 Watts
23	RPT-1	462.5500/467.5500	5 Watts
24	RPT-2	462.5750/467.5750	5 Watts
25	RPT-3	462.6000/467.6000	5 Watts
26	RPT-4	462.6250/467.6250	5 Watts
27	RPT-5	462.6500/467.6500	5 Watts
28	RPT-6	462.6750/467.6750	5 Watts
29	RPT-7	462.7000/467.7000	5 Watts
30	RPT-8	462.7250/467.7250	5 Watts

Here is a simplified diagram that shows GMRS channels and repeaters:

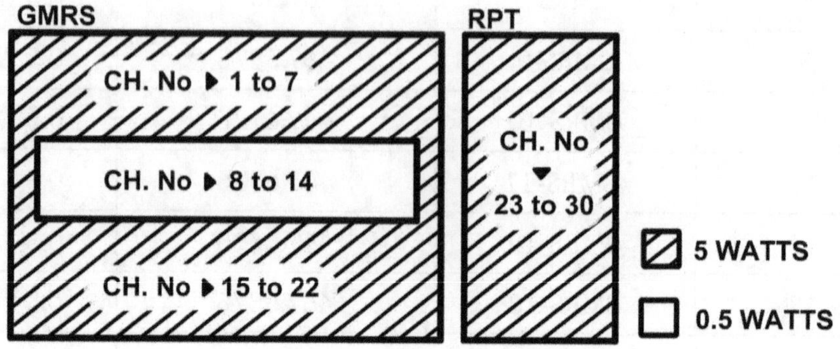

PRE-PROGRAMED CHANNELS DIAGRAM

Well, we've been able to look into the incredible flexibility unlocked through tailored programming! We were able to transform this formerly intimidating device into a versatile multi-channel communications hub optimized for diverse needs. We'll now go ahead to cover how to improve range and audio quality in the next chapter.

CHAPTER 6: ENHANCING RANGE & AUDIO QUALITY

This chapter focuses on maximizing your Baofeng's effective reach for quality audio, no matter the digital chaos surrounding modern urban environments.

Whether coordinating volunteer events spanning stadium parking lots or marshaling backcountry search parties scattered across mountainsides, optimizing hardware, and accessories tailored to usage needs, empowers success.

We will thoroughly cover proven techniques boosting transmit distance, decoding muffled signals, bypassing interference, and configuring programmable settings to unlock intelligible performance near and far.

Let's conquer winnable basics like antenna orientation/tuning, terrain/structure impacts, interference types, and hardware tweaks before exploring expensive amplifier upgrades or high-gain directional antenna arrays used by radio enthusiasts seeking extreme reception feats!

6.1 Improving Range & Clarity

Getting clear radio signals to reach a widespread group over large areas is crucial in crises. However, increasing the distance your radio broadcast takes balancing fancy gear prices and complexity.

This section discusses things that increase range, such as boosting signal power, improving antennas to focus direction, finding clear, direct paths for messages using high places like hills or rooftops, and more range-extending options.

We'll compare upgrade choices that help simply improve talk distance without getting overwhelmed by confusing menus or gear overload. Follow along as we maximize what these affordable Chinese radios can do through intelligent, easy add-ons tailored to your needs!

Checking Your Radio's Range

You'll need to take the radios out of the box with included antennas and batteries to see how far apart you and a team member can communicate clearly.

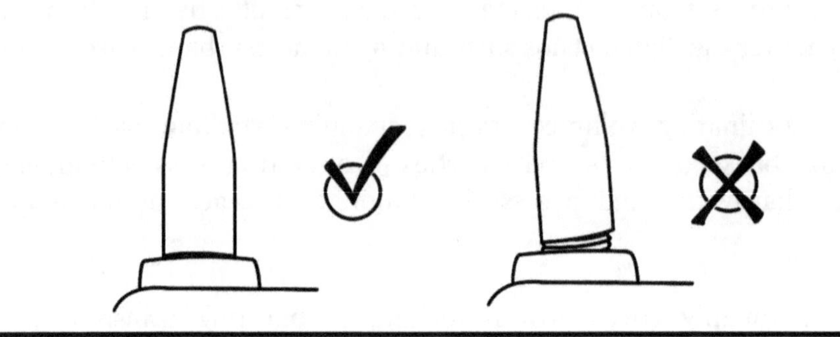

MAKE SURE THE ANTENNA IS INSTALLED CORRECTLY

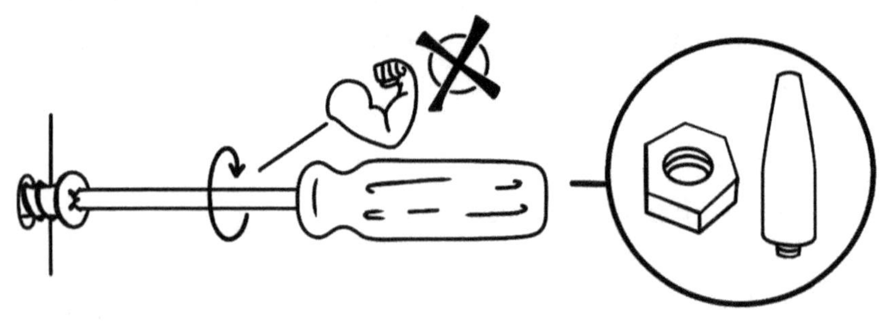

DON'T APPLY A GREATER FORCE ON THE ANTENNA ROTATION

Run a simple test by having your partner walk away slowly while you both talk. At some point, static cuts in, and voices become unclear. Note the distance where reception starts fading.

This real-world test gives the maximum reliable talk range using standard equipment. If your needs allow groups to stay within the tested distance, you don't need upgrades. Save your money!

But if team members often roam farther than the initial static point, consider range boost options—improved antennas, better positioning, more broadcast power, etc. If possible, only spend money selectively addressing defined range shortcomings shown in testing.

Turning Up Transmit Power

Turning up the signal strength your radio uses to broadcast lets messages transmit much farther—it's like yelling louder to be heard across a bigger room.

But one downside is it consumes batteries much faster as loud as your transmission power gets. So, if you make the range far by maximizing power, be ready to swap backup batteries more often, or they'll go dead quicker.

One option is choosing extended-life battery packs rated for longer talk time. That way, even at total volume, the better batteries won't die faster.

However, spare regular batteries should also be packed as a contingency since even the best cells eventually drain if transmitting total loudness nonstop. Balance range must be determined by how often you're willing to recharge/swap as power capacity allows.

Clear Lines of Sight Help Signals

Walkie-talkie messages ride on radio waves, moving straight out like beams from a flashlight or laser pointer. If trees, hills, or buildings block the straight line between you and your radio partner, signals get interrupted like shining a flashlight against obstacles.

But just like towering stadium lights reach far, casting downward, raising antennas on high poles, in raised clearings, or on top mountains shoots signals out much farther before any obstacle blocks the "view".

Keeping your radio and partner's radio antenna clear and on a straight-line path goes a long way toward maximum range. That's why elevating either end of the chat by getting on top of trucks or roofs extends reliable talk coverage over more ground.

The key idea is that uninterrupted straight visibility between radio antennas allows the signals to propagate much farther, clearing more obstacles.

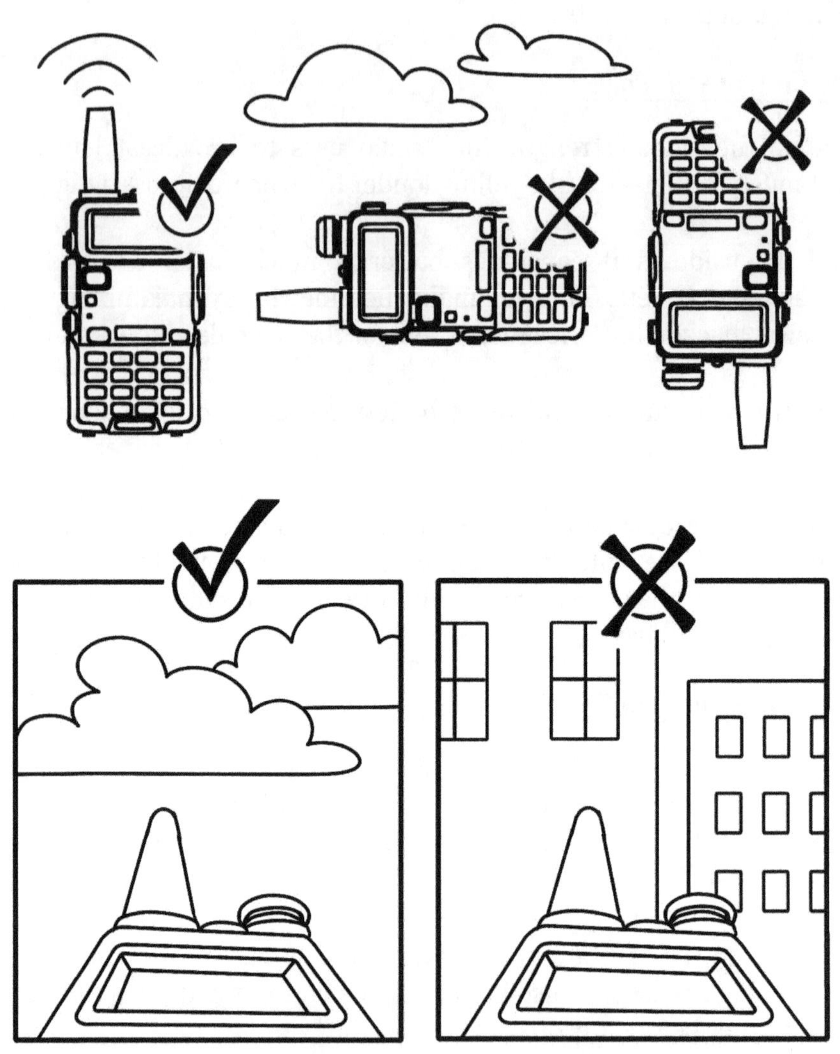

THE PROPER RADIO POSITION

Getting Better Range from Your Antenna

Attaching more giant external antennas can boost range over the small antennas that come with the radios. Upgraded antennas focus the radio signals like a beam rather than spraying randomly.

It's like adding a megaphone compared to just yelling. But more giant antennas cost more, can break easier, and may catch on stuff when carrying, so choose wisely.

For many needs, an affordable mid-size antenna attachment offers significant improvement without going overboard on gigantic trucker gear. Test in your actual conditions, starting small.

Before buying the expensive heavy-duty options, see if an essential external antenna addition solves the distance issue. Rather than overspending upfront, closely check what the extra mileage tests reveal from incremental upgrades. The intelligent antenna picks these pocket radios!

6.2 Antenna Selection & Tuning

Baofeng radios pack a substantial range into compact packages, but the included antennas are basic, impacting reception, as noted earlier. However, affordable third-party antenna upgrades can quickly compensate for the antenna shortcomings that come standard.

The antenna shapes the radio waves that are broadcast from the radio. It affects how far the signals can reach.

There are many antenna options - some are just a few inches long, while others are several feet long. Each antenna is suited for different needs.

So, it's essential to carefully choose the right type of antenna for your needs before buying one. Picking the wrong antenna can limit your range or reception.

Longer is not automatically better! Evaluating key metrics like frequency range, directionality, and gain patterns prevents squandering cash on incompatible or under/overkill units, ultimately degrading functionality through mismatched impedances.

Amateur installations also risk transmitted signal leakage, which can degrade local reception without proper impedance matching and filtering. But when judiciously paired with proper 50-ohm loads, massive range improvements await!

Let's break down specifications toward ideal antenna mating.

Frequency Range

VHF radios operate between 136 and 174MHz, while UHF typically uses 400 and 520MHz. So, ensure chosen antennas explicitly support designated commercial 2-way radio bands or advertised amateur 70cm wavelengths.

Directional or Omnidirectional Radiation Patterns

Omnidirectional whip antennas evenly distribute signals across 360 degrees, suiting scattered user mobility. Directional beam antennas focus energy pools, intensifying strength along particular axes at the cost of dead zones off boresight. Evaluate expected caller positions before unthinkingly purchasing!

Gain Specifications

Antenna gain indicates energy-focusing efficiency with decibels representing 2-3x power multiples. Too many risks overpowering nearby receivers, while too little leaves marginal walkie-talkie connections scratchy. Consider 5-6dB models optimizing moderate boosts limiting distortion reception issues.

Also, consider functionality when shopping for replacement antennas:

Telescoping models: Collapse compactly when portability matters most. But match length to intended frequencies for optimized performance.

Base Loaded or Helical antennas: Also shrink but use coiled wires, allowing broader reception across full bands.

Signal Sticks offer flexible, slim blades under a foot yet capture admirable gain. They can be tossed in backpacks for ultra-light backup.

Multi-band Nagoya models: Cater to both VHF and UHF in a single rubber-sheathed antenna that is barely longer than the handheld height. Dual-band coverage!

5/8 or 1/4 wavelength premium fiberglass models deliver reliable 360 coverage for fixed base station installations without unwieldy length. Consider weather protection and solid mounting to extract the entire gain potential leveraging height/locations.

And if aesthetics permit, absolutely utilize exterior rooftop antennas whenever feasible! Even a primary whip antenna mounted atop a 30-floor skyscraper offers exceptional reception, offering 100X+ range-stretching gains to handhelds many miles distant. Exploit height!

Some tuning trickery improving reception:

Carefully trim antennas by centimeters, progressively checking for increased signal clarity as you establish peak length dialing into your desired transmit frequency. Chances are factory antennas cut for generic bands, not YOUR programmed channels.

Also, gently bend base sections to seek alignment, delivering the cleanest reception/transmission by minimally obstructing the internal antenna path to board electronics. You want the external antenna broadcasting signals absorbed close-coupled to the circuitry without veering off course!

With quality antennas properly aligned/tuned, remarkable improvements transform formerly spotty connections into booming exchanges spanning impressive distances - especially when height enters the equation!

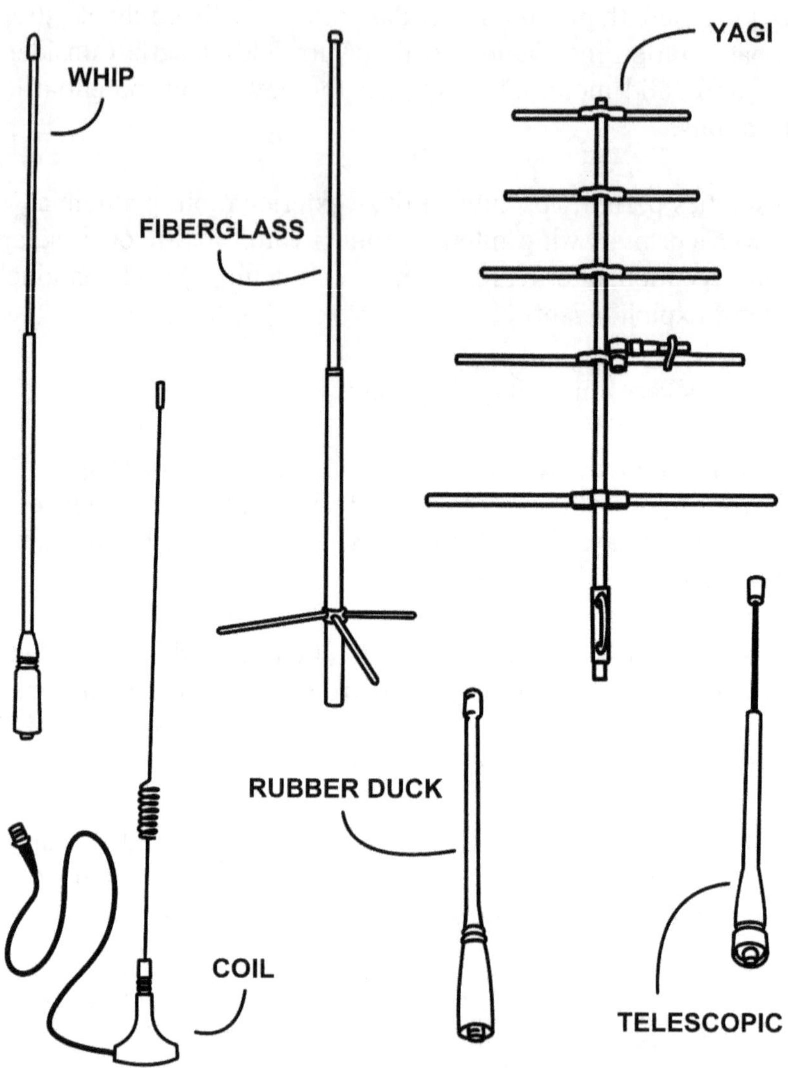

DIFFERENT ANTENNA TYPES

6.3 Mitigating Interference

Even optimized configurations still contend with digital chaos flooding crowded bands. Whether bleed over from adjacent blend users, background solar radiation, or device self-noise from insufficient filtering, unwanted contamination cuts signal integrity.

Combat crony transmitters are stomping conversations by configuring sub-channels and applying selective digital signal processing.

CTCSS/DCS Codes

Transmitting encoded squelch tones quiets receiving speakers until matching codes are detected, preventing unwanted eavesdropping. This also silences noise, and lacking codes keeps channels more precise.

Narrow Bandwidth

Narrowing your scanner's listening range to focus in—Set your scanner to listen to only a small range of 0.4 kHz radio waves instead of extra-wide 5 kHz waves. This tunes out noisy signals around the words people are saying, so your scanner picks up the message more clearly without the junk sounds blended in.

Audio Limiting

Overdriving receiver circuits also invite distortion. Reining excessive amplification prevents overload and maintains high-fidelity scanner quality. Let clear signals pass unadulterated by tighter constraints.

And if all else fails, manually notch filter persistent interference or creep, transmitting offsets gradually to new, unused neighboring frequencies. Radio spectrum space remains surprisingly vast if you know where to prospect!

Well, that concludes our graduate course on unlocking extended radio reach! We explored antenna types and alignment tricks, propagation mechanics leveraging line-of-sight and building penetration traits across bands, signal contaminants plaguing crowded channels, and hardware/software filtration techniques combating disruption.

CHAPTER 7: EMERGENCY COMMUNICATION STRATEGIES

Being prepared with solid radio know-how when emergencies strike makes a difference in quickly coordinating responses and staying informed. When critical situations emerge, explore intelligent strategies to get the most from these Baofeng walkies.

I'll explain handy built-in features that shine when catastrophe hits: efficiently pre-programming regional emergency frequencies, running practice drills to build effectiveness, and even bridging into wider disaster networks if phones or the internet fail.

Having backup plans for sufficient power, alternate charging sources, contingency hardware, and making quick repairs guarantees continuous radio contact when every message proves vital. Safety starts with strategy!

7.1 Emergency Communication Planning

Carefully determine likely emergency scenarios based on geographic region, prevailing weather threats, group excursion locations, local infrastructure robustness, and incident history trends.

Does extended drought elevate wildfire risk? Do remote camping/hunting areas lack nimble backup options if injuries occur? Have recent tornadoes revealed repeating blindspots in municipal early warning coverage through after-action findings?

Such realism steers prudent contingency planning for when—not if—the next eventually strikes. Tabletop thought exercises imagining scenarios prove outstanding training, revealing priority frequencies and tactical response improvements required. Stage "pretend incidents" and practice protocols across the community!

Pay attention to built-in radio safety features like LED flashlights, loud audible alarms, and weather band receivers, which offer extended value in disasters. Assess and enable all available capabilities upfront, thinking broadly!

7.2 Connecting with Emergency Networks

Public agencies publish designated frequencies supporting health/welfare coordination for qualified volunteer organizations during incidents. Having those channels pre-programmed means instant access if mobile networks get overwhelmed.

Additionally, research popular ham emergency frequencies where mobilized auxiliary communicators converge, establishing self-organized relief coordination. Even without licenses, monitoring the evolving Temporary Amateur Radio Service provides invaluable situational awareness as events unfold.

Web research identifies the exact MHz numbers designated for your county/ state. Record these channel details and label CTCSS tones centrally for easy reference if critical infrastructure later sustains damage. Don't rely solely on mobile connectivity to convenient digital frequency databases when racing response clocks!

7.3 Establishing Redundant Backup Systems

The difference between a rescuer radio dying mid-mission versus sustained around-the-clock ready capability is vital in establishing command chains. We must guarantee electrical independence and resilience!

Car chargers allow topping off from vehicles while portable solar panels recharge handhelds indefinitely. Always packing ample spare cell packs provides physical redundancy as well.

Consider homemade Faraday cage shielding, stored kits in metal boxes or mesh bags, blocking damaging solar flares that disrupt global comms. That sounds extreme, but blackouts teach hard lessons when stranded!

Also, secure backup antenna equipment and program cables permit rapid field repairs or reconfigurations no matter the damage sustained in chaos. Refrain from assuming anything about supporting infrastructure when planning for worst-cast contingencies!

7.4 Training & Roaming Considerations

Stage repetitive practice drills on anticipated channels, confirming adequate range from secured rally points or critical infrastructure sites. Understand that familiarity builds operational readiness if working while scared or frantic when crises strike!

Listen to regional first responder exchanges during their regular non-emergencies activations. This better calibrates your antenna tuning to local chatter signatures and protocols. Program all observed primary dispatch frequencies, ensuring seamless interoperability when lives hang in the balance!

Consider roaming privileges bordering statehoods required. Some accept credentials universally, while others enforce exemptions and additional qualifications. Limited windows of external specialist operation may provide sufficient support nonetheless. Just beware of airtime rules as events unfold dynamically!

7.5 Cutting Through the Chaos

Even with good gear and channel lists, sending messages is still challenging in chaotic situations. Loud noise buries signals. We need eye-catching plays to punch through the madness and be noticed when things get wild...

Give every team member a unique call sign. Now you can call out specifics like "@Eagle25" even when many voices are shouting. This lets your message cut through directly to the right person instead of getting lost in the noise. No more unclear yelling without knowing who should respond!

Say "Urgent! Listen up!" first when sending important info. Repeating these attention-grabbing words helps slice through loud chaos, tuning people's ears to actively hone in on your voice before telling them the emergency play.

Also, leaders should be put on open speaker channels. When important folks start talking instantly, their voices immediately blast out to entire teams. Everyone hears the big news when leaders begin sharing what to do. No claiming, "Oops, didn't get the message!"

Use unique warning beeps before sending significant alerts. When bad stuff goes down, all the channels fill with so much yelling no one can listen. Teams try making their plans, talking over each other. It becomes a mess of overlapping noise! Make your emergency update blare out by playing unpleasant blaring tones first. That'll hush them up!

After practicing these tips to get messages through, if stuff goes south and drilling them nonstop, you can stay chill even when significant trouble strikes. Keep people talking even with systems down and too much demand! What we covered unlocks a solid plan for when bad scenarios unfold, and trades get chaotic. Just stick to the plan, and you'll get this!

7.6 Testing Radio Readiness

Burying our heads, assuming the programming and policies we've set will automatically activate without hesitation when lives on the line lead to disaster. Taking readiness confidence as a given invites complacency. We must proactively train responses to second-nature levels through relentless surprise examinations exposing weak links.

Let's implement a gauntlet readiness exercise series preparing our team coordination reflexes to near automated levels:

Random Radio Checks

At completely unpredictable intervals, interject emergency channel hails, demanding immediate check-ins, status updates, resource requests, and next-step instructions. Regardless of activities in progress, these interrupted simulated crisis alerts are prioritized over existing engagements. Even minor delays or confusion in logging positions and reactions could prove catastrophic in actual incidents.

Distraction Combat Scenarios

Once fundamental radio check capabilities are benchmarked reliably, you should induce intentional distraction elements while processing emergency instructions. You might need to recite critical medical treatment steps precisely while your radio blasts off-track casual banter in one ear. Getting comfortably immersed in your response roles despite surrounding chaos accelerates realism.

Equipment Failure Chains

Finally, no proper menu configuration prevents batteries, antennas, or hardware from potential damage mid-crisis. We must learn to bridge communication chains absent modern tools. You should randomly revoke individual radio access remotely. This forces the relaying of messages verbally from member to member when direct contact is impossible. Adapting without hesitation keeps the mission continuing forward.

By subjecting our emergency readiness assumptions to volatile surprise evaluations, we shift from academic hypotheticals to instincts etched through relentless exposure therapy.

CHAPTER 8: ENCRYPTION AND SECURE COMMUNICATION

Encryption is crucial to prevent unauthorized eavesdropping when transmitting private or sensitive information over the airwaves. This chapter thoroughly covers configuring your Baofeng radio to scramble communications using analog and digital encryption protocols. We'll cover encryption concepts, step-by-step configuring instructions, scrambler accessory recommendations, and even tips on generating ultra-secure passphrase codes to future-proof your wireless security against even sophisticated attacks with today's consumer two-way radios.

8.1 Encryption Overview

At a basic level, encryption converts a radio signal containing everyday human speech into a modified, digitally coded format that sounds like noise to listeners lacking the proper decryption key.

It's like disguising your voice with a secret spy decoder ring so only fellow agents can understand! The technology protects confidential chatter from curious ears by altering signal patterns transmitted.

Encrypted signals ride on standard radio frequencies like any other transmission. However, advanced encoding algorithms mathematically transform voice data into unique sequences only intended receivers can decrypt, keeping conversations private.

Analog vs Digital Encryption

There are two primary methods for encrypting radio communications:

Analog Encryption: Converting voice to varying electrical signal patterns using primary audio effects before broadcasting as an analog signal. Examples include pitch manipulation, bandpass filtering, noise injection, and inversion. While analog techniques are simple to implement, digital encryption tends to be more robust.

Digital Encryption: Digitizing voice first into binary data for encryption through complex mathematical algorithms before transmitting noise-like modulated packets. The information stays far more protected using state-of-the-art techniques like AES or SHA cryptographic schemas by converting the entire signal chain to digitally secured binary before encrypting/transmitting/decoding. However, digital requires more processing overhead.

Baofeng Radios typically rely more on analog scrambling methods or basic digital protocols to balance overhead versus security for largely unregulated civilian use. First, an encryption primer must be used before assessing available options.

Encryption Keys Concept

The critical ingredient allowing encrypted communications is having the correct decryption key! Unique encryption keys are generated and shared with authorized radio operators to reconstitute scrambled messages.

Keys contain mathematical instruction sets for decoding. Without access to associated keys, encrypted signals remain indecipherable noise.

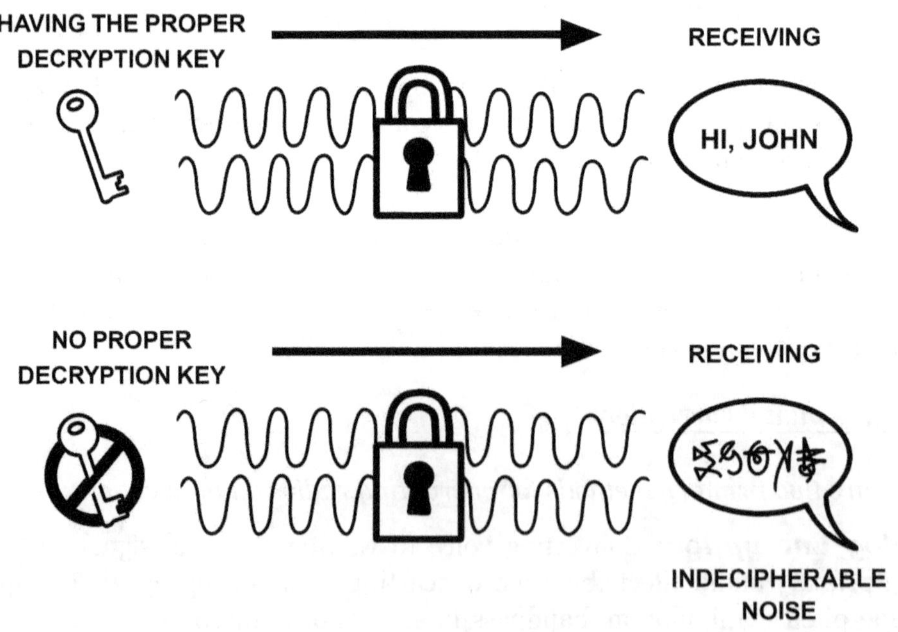

ENCRYPTED CHAT

There are two general approaches used:

Symmetric Encryption: Symmetric Encryption utilizes a single Shared Key between users. This standard passphrase string gets distributed (hopefully secretly) among your group. Units mutually decode encrypted transmissions as long as all members enter the identical key into their radio config. Any outsiders without access to that key just hear garbled noise and cannot unscramble messages.

Asymmetric Encryption: Asymmetric Encryption relies on Public Key / Private Key pairs. Your radio can only access the Public Key to encode messages before transmitting encrypted packets to others. However, the intended receiving party holds the corresponding Private Key capable of decrypting and recovering the original pre-encrypted data through additional processing. This split method provides another layer of verification but gets complex for most basic walkie-talkie needs.

Baofeng Radios focuses more on Symmetric Encryption, given hardware constraints and civilian ownership use cases that prioritize simplicity over high-threat security.

8.2 Configuring Privacy Codes

The first step enables digitally encoding communication channels for basic scrambling between other radios containing your codes. Here is the quick process to secretly configure your radio cluster for secure transmission between members:

Make a random Mix Code number string to identify your encryption group across units uniquely. For example: "835272".

In CHIRP programming software, enter your unique Mix Code under Settings > Digital/Analog > Talk Group ID field.

Under Settings > Digital/Analog settings, choose the desired Analog Encryption option:

"Inverted" simply inverts the analog signal phase using minimal processing.

"Scrambled" more aggressively distorts the audio signal for better protection.

Ensure all distributed radios have matching Talk Group ID fields and Encryption settings.

Once the Talk Group ID contains your custom string, encryption activates between units that share that code phrase. Verify radios match in your fleet!

Advanced Analog Protection

Beyond basics, enthusiasts pursue analog encryption through external accessories like voice scramblers or distortion pedals for even stronger obfuscation effects securing channels.

Scrambler Units act like outboard signal processors, sitting between transceivers and audio sources, actively transforming and distorting. Professional variants leverage rapid rolling code generation and custom algorithm sets for potent Scrambling that exceeds baseline radio capabilities. Expect to budget $200+ for dedicated hardware, though!

Audio FX Pedals offer creative analog protection through intentional manipulation, such as pitch/tone changes, resonant peaks, noise injection, sample stuttering, etc. They are more accessible for home tinkerers and musicians! Carefully tune effect settings, training ears to mentally "decode" altered audio. It gets exciting results!

Of course, analog methods still have limitations for protecting highly confidential exchanges. But casual privacy can be boosted through accessible, affordable means!

8.3 Digital Encryption Options

For even deeper defense, by applying accurate encryption algorithms like AES, DES, or RC4 against advanced threats, we enter the domain of digital encryption protocols that are purpose-built for secure transmissions.

Unfortunately, Baofeng models lack native hardware encryption engines found on military or public safety handsets for crunching intensive algorithms. We need external digital assistance!

Let's examine solutions commonly paired with Baofeng hardware extending security:

ADI Encryption Devices allow users to connect dedicated AES-engine accessories via external headset ports. Encrypted digital voice packets transmit securely, allowing only other ADID units to decrypt streams into understandable audio using pre-loaded keys. Expect ~$300 investments protecting each radio endpoint but rock-solid security.

Third-party apps like Zello offer app-based encryption options when wirelessly tethering transceivers to smartphones. Scrambled packets relay securely between app instances using robust 256-bit algorithms, shielding messages from prying ears or tools like Stingrays. This is convenient but drains phone batteries quicker!

Custom Firmware and cables support augmenting stock handheld systems with open-source software foundations, enabling enhanced native encryption routines on enhanced radio hardware. OpenGD77, JumboSpot, and others unlock advanced functionality through DIY modifications but require technical skill sets.

So, digital encryption beckons for fortifying communication beyond the scrambling basics built into most Baofengs. Determine security priorities by balancing convenience, expertise, and cash outlays, tailoring the ideal solution per your unique requirements. Plenty of options now exist, extending protection far beyond older analog stopgaps!

8.4 Generating Secure Encryption Passphrases

All encryption hinges on the secrecy encrypting keys introduced. Weak passphrases succumb to common dictionary attacks. But cumbersome overcomplication backfires operationally through countless garbled mistypes and retries.

Let's review tactics crafting simple but strong encryption keys proven to protect exchanges while remaining realistically recallable for hectic real-world radio troops:

Length First: The most extended key surviving intact in memories wins. Even essential words strung 8+ terms prove harder to crack than clever quips falling short.

Easy Flow: Choose easily visualized passphrases or stories that make coherent sense. Recall the sequence naturally by chunking related concepts. Random gibberish lacks mental hooks.

Mixed Cases: Capitalizing letters Mid-Word adds variation, foiling frequency analysis seeking common vocabulary or repeated governmental names as weaknesses. Break expectations!

Number Substitution: Replace common letter pairs with associated numbers or symbols. "E + L" becomes "3 + !"; "A + N" converts to "@ + ^". Match substitutions consistently across your group.

Favor Newer Terms: Preferentially pick newer pop culture names and vocabulary not present in shared password dictionary lists assembled from past leaks. Keeps guesses behind the times!

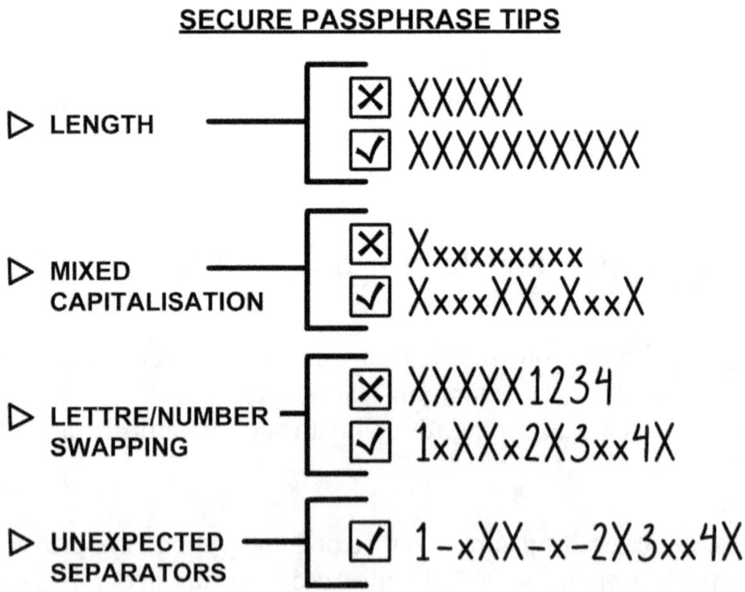

With these guiding principles, even amateur radio networks stand reasonable chances of repelling external monitoring given modest threat models associated with Baofeng walkie-talkie use cases.

Advanced perpetual rolling cipher generators offer near-NSA Grade impenetrability safeguarding Special Forces operations. But common-sense conventions already outlined thwart bored neighborhood kids randomly scanning through frequencies hoping for clues about that weekend keg party location! With some creative flair, your communications stay off the radar without needing master's degrees to decipher mathematically optimized encryption schemas.

The foundational layer is establishing shared secret passphrases across your distributed radio network. Once your team internalizes the authorized talk group ID, casually consulting crib sheets, encryption fortifies exchanges using built-in software settings or upgraded analog/digital accessories detailed earlier. Now, proximate signals stand little chance of interfering with intended communications inside your circle!

CHAPTER 9: PERFORMANCE ENHANCEMENTS

Let's unlock the full potential of these pocket radios! This chapter dives into accessories and tricks that push range, clarity, and overall capabilities to the limits.

9.1 Antennas for Improved Range

The short, fixed antennas on Baofengs work fine enough for essential uses. However, larger external antennas can massively boost signal reach for further communication. It's one of the most accessible and affordable ways to upgrade performance.

Think of it as upgrading a primary bunny-ears TV antenna on an ancient tube set to the modern digital antenna on a flatscreen pulling in channels clearly from 50+ miles away. The same idea applies to radios!

There are a few antenna qualities to evaluate:

Gain: Measured in dB, this tells how much an antenna amplifies and focuses signal strength in specific directions. Higher gain number = louder signal broadcast and received. But watch height limits for your radio's power.

Direction: Omnidirectional antennas send signals evenly all around. Directional ones focus energy like a beam in specific directions only. Choose whether you must reach people spread everywhere or just in one target area.

Frequency: Match antenna range to the MHz frequencies your radio uses, like VHF or UHF bands. An antenna rated for 450-470 MHz works better for UHF Baofeng models than a broader 150-520 MHz antenna. Closer frequency matching concentrates performance.

Type: Form impacts function! Whip, coil, base, and fiberglass antennas all act uniquely. Consider size, extension, flexibility, and mounting needs, choosing the right physical antenna style.

Here are a few great starter antennas to consider that won't break the bank:

Nagoya NA-771: This 15.6-inch flexible coil antenna is rated for the 136-174MHz and 400-520MHz bands. It covers both VHF and UHF in one rugged unit. It improves reception and broadcast range over a stock antenna and is under $20.

Authentic Signal Stick: Slim, flexible antenna made from strong 3003 aluminum alloy rated for comprehensive frequency coverage. At 16.5 inches, it packs excellent gain while quickly slipping into pockets. Around $20 bucks makes this a top choice for preppers and hikers needing portability.

Ed Fong DBJ-1 Roll-Up: Unique roll-up design packs down tiny but unfurls to 19 inches for excellent portable performance across 137-520MHz frequencies—rugged, flexible steel construction. Attach to metal surfaces for an even more considerable gain boost! Just $30.

Nagoya UT-72 Magnet Mount Base: This base converts existing antennas for magnetic mounting on vehicle roofs or metal surfaces. It makes mobile use easy by leveraging height for an expanded range. It safely holds 1/4-wave antennas under 54 inches long. This is a $35 value upgrade.

These cover some of the best mid-range antennas, offering significant improvements for average users without spending big bucks. Feel free to go fancier, but test models in your actual conditions before buying the $100+ showcase units. An affordable antenna carefully matched to your exact radio and use case may work just as well or better!

9.2 Accessories for Clarity & Capabilities

Beyond just signal reach, additional accessories further increase the effectiveness of applying your radio in specialized scenarios:

Headsets: Headphones with integrated boom mics allow discrete radio monitoring and communication. They are necessary for security teams, event staff coordination, and any uses needing a hands-free operation or private listening. Choose comfortable over-ear versus earbud styles based on use duration. Prices range from $20 for basic sets to $200+ for heavy-duty noise-cancelling versions. USB programming cables allow quick in-field reconfiguration and firmware updates leveraging laptop access. It is far easier than manual keypads for tweaking settings between events or across locations. Choose the correct plugs for your radio. It's around $10 per cable; having spares for teams is good.

Speaker Mics: These convenient clip-on microphones with an integrated speaker free up hands. They allow you to listen and talk while the radio can be carried out of the way on a belt or vest. Ideal for continuous all-day use, they cost $25-$50.

Batteries: Always pack spare charged batteries for all-day readiness! An extra battery or two in go-bags guarantees you won't get stranded mid-mission. Opt for extra high-capacity batteries up to 5000mAh whenever possible over standard 1800mAh ones.

Cases: Protect your radio from drops, debris, and weather with a sturdy case. Chest harness straps or Molle attachments allow efficient carry during movement or observation posts. Never rely on loose radio pockets! $15-$50 for serious protection.

By now, you're experienced in picking the right gear. Evaluate your ideal setup, then accessorize thoughtfully!

9.3 Boosting Signal Strength & Quality

A few key settings tweak how Baofeng radios broadcast and receive signals in ways that impact overall link quality and distance:

Transmit Power: Move from low 1-watt transmission to 8 watts for max broadcast power. However, a shorter range saves battery duration. Five watts balance performance for most—access in menu options.

Squelch Level: Adjust the threshold to mute weak signals or noise while allowing clear nearby communications. This prevents constant static. If needed, a lower threshold opens reception to more distant signals.

Mic Gain: Turn up the mic volume, picking up your voice for vital outgoing signal clarity. But don't overdo it; it will transmit distortion if it's too hot.

Receive Boost: This amplifies the incoming signal gain for increased loudness and rango DNA, allowing marginal stations to punch through. Don't blow your ears out, though!

These four adjustments optimize two-way exchanges. Ramp them while testing with partners until both send and receive sound crisp and clear, even at a range near the threshold where background static starts creeping in. That dial-in yields ideal calibrated setting combinations customized for your actual radio equipment and intended use locations.

The other massive benefit beyond fine-tuning voice signals is enabling solid data transmission utilizing the radio's AFSK capability. With tuned performance, Baofeng becomes capable of surprisingly decent throughput for basic field sensors and embedded systems connectivity.

Turn Your Baofeng into a Data Radio!

With great care in configuration, Baofeng radios transform into capable AFSK modem platforms for linking remote field equipment back to operator sites.

This allows telemetry gathering for things like:

- Environmental sensors.
- Agricultural equipment.
- Weather and seismic stations.
- Trail cameras.
- Water vessel autopilots.
- Payload balloons and drones.
- Underground infrastructure monitors.
- Experimental robots.o

...plus endless other monitoring devices!

The key is utilizing audio tones shifted between predefined high and low frequencies to digitally encode 1s and 0s in the radio signal.

On the remote device end, a microcontroller translates sensor data into this AFSK-modulated audio sent over the air by the Baofeng transceiver.

At the receiving Baofeng, this tone sequence gets demodulated back into binary packets containing the original sensor payload data.

With optimized antenna tuning and performance configs, surprisingly decent throughput reaches 300 to 600+ bps, matching basic GPS trackers and simple serial devices. Though not lightning fast, this opens excellent possibilities!

For example, hikers could deploy simple weather stations deep in the backcountry, transmitting snapshots back to base, or a research team could monitor seismic events from remote regions.

The only limiting factors are power availability at remote device points and baseline radio range. But impressive distances are covered by incrementally hopping data using a mesh of Baofeng nodes!

This little trick unlocks substantial capabilities many overlook for Baofengs as simple walkie-talkies. With the proper optimizations, flexibility expands using existing infrastructure in innovative ways!

We've covered a lot here. This chapter has enhancement concepts that push these wallet-friendly radios to new heights! We explored antenna upgrades that reach farther and punch through obstacles. Accessories like headsets, mics, and batteries enable superior field use. Performance tuning adapts signals optimally for voice or even remote data links.

The key is carefully choosing improvements that align with YOUR intended usage, environment, and budget. An expensive high-gain antenna may look impressive but prove overkill for short-range urban teams. Complex accessories add capability yet also introduce fragility and training demands.

Evaluate options thoughtfully and test incrementally. Substantial performance gains are unlocked for less than $100 in upgrades, turning bargain Baofeng walkie-talkies into specialized powerhouse solutions! You now have deep knowledge of equipping these radios precisely as your application requires.

CHAPTER 10: TROUBLESHOOTING AND MAINTENANCE

Radios take a beating when used in demanding real-world conditions. Let's cover common problems arising during use and preventative care, ensuring your Baofeng lasts as long as needed.

10.1 Common Issues and Solutions

Even quality radios act up occasionally. Here are some typical problems and fixes to get you quickly back communicating:

Power Problems

If your radio won't turn on, remove and re-insert the batteries. This resets the connections. Then, check if the batteries are dead—replace them if they have no charge left. Fully recharge depleted batteries to restore power.

If it is still not working, dirt might be blocking essential contacts. Gently clean the battery contacts using alcohol and cotton swabs. Be very careful, though! If problems continue, you may need manufacturer service for internal faults.

Buttons Not Working

First, check if the keypad lock button on top is enabled—a lock icon will show on the screen. Tap it again to turn off the lock if it is on. Now, try the buttons again.

If the unit is adequately unlocked but the buttons still don't function, try a manual reset by swiftly powering the unit on and off. This recycles components, allowing recovery from any temporary glitches.

Consider a firmware update or factory reset for chronic issues to erase corrupt settings. If the problem is hardware-related, contact the manufacturer to initiate a replacement/warranty claim, given the continuous unresponsive keys.

No/Low Audio

Quiet reception and transmission often stem from the exact causes. Check essential factors like volume and squelch levels first before hardware assumptions.

Inspect speakers and microphone ports for obstructions like dust-blocking sound. Clean gently using compressed air.

An accidental setting like voice prompts disabling audio feedback may be enabled - scan settings to revert. Otherwise, contact the manufacturer to diagnose potential component failures needing repair.

Short Range/Weak Signals

If distances were previously longer, evaluate the antenna and input/output port integrity first. Also, check for damaged connectors or antennas in the slash range. Inspect carefully and replace if needed.

Also, scan areas for new sources of interference that may prevent clear communication. Adjust frequencies or coding settings to overcome ambient signal competition if simple repositioning doesn't solve range degradation.

Battery Issues

Batteries eventually degrade with heavy use. If operating times gradually shorten over their lifespan, replacing cells restores total power capacity. Also, ensure contacts remain clean.

If batteries no longer hold a charge, improper charging will likely damage cells. Safely dispose of/recycle to avoid leakage hazards and follow proper charging procedures with new batteries. Only use compatible chargers to maximize battery lifespan and performance.

Water Exposure Issues

Moisture in electronics never mixes well! If radios seem erratic or unresponsive following wet weather use or accidental submersion, immediately remove power sources and disassemble them to dry interior components thoroughly. The function may recover after a couple of days of drying time. But long-term damage risks remain, so factor in replacement cost/availability just in case. Lesson learned about weather protection!

Programming Problems

Issues with reading/writing frequencies and settings often trace back to cable connections or software glitches corrupting data. Reboot computers, temporarily use a simplified manual programming process, and retry with fresh USB cables to rule out system-related factors.

For settings randomly resetting/deleting or buttons not matching expected programmed behaviors, a factory reset wiping settings may resolve software corruption. Backup data beforehand!

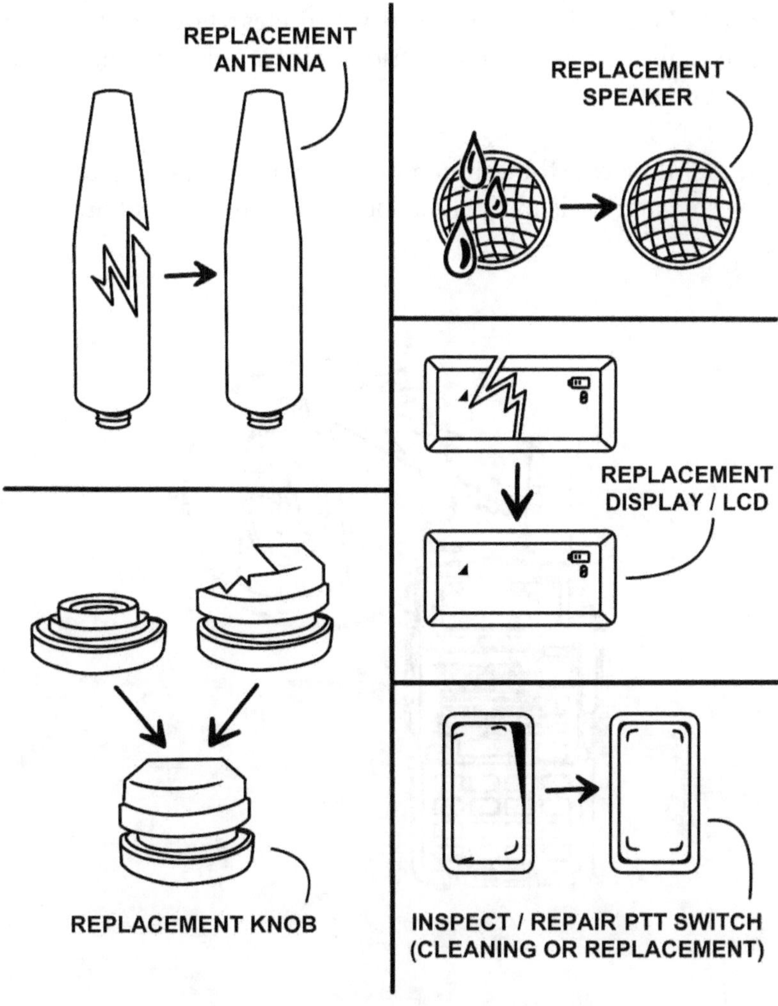

COMPONENTS THAT FREQUENTLY FAIL AFTER DROPS OR DAMAGE

10.2 Cleaning and Maintenance

Preventing issues through proactive steps means less downtime later! Let's cover best practices for maximizing your investment lifespan through proper care:

Antenna: Regularly inspect bases for looseness and test retention strength, twisting gently. Overtightening risks internal damage, so find the sweet spot to balance connection integrity with excess torque.

Buttons and Ports: Use compressed air to clear any accumulated dust or debris on external controls and connection points. This prevents oxidation and intermittent contact.

Housings: Gently clean the outer radio body using microfiber cloths lightly dampened with water or isopropyl alcohol when dirty. Avoid abrasives risking scratches.

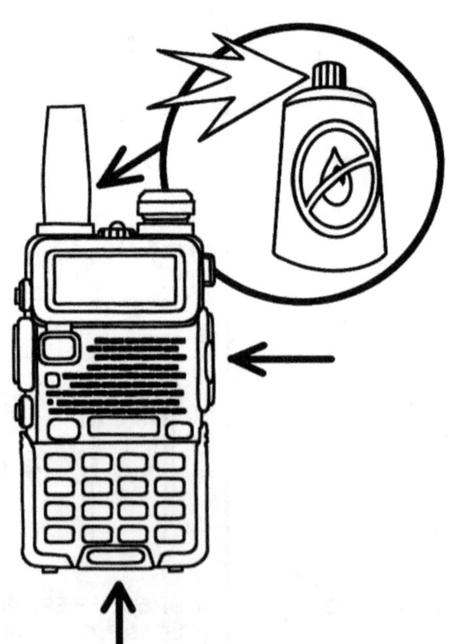

USE COMPRESSED AIR TO CLEAN PORTS.
AVOID ISOPROPYL ALCOHOL/WATER CONTACT

Batteries: When not using rechargeable battery packs for extended periods, discharge them to around 50% charge before storage. Remove them from radios and store them in a cool, dry place to maximize their lifespan.

Accessories: Make sure to disconnect headsets/mics after use to avoid strain on ports over long-term usage. Collapse antennas when not needed for transport. Inspect all parts before/after missions.

Periodic Resets: Employ master resetting to factory conditions. A fresh software start every few months clears any latent performance gremlins missed during standard operation.

Careful Care Pays Off!

Consistently practicing preventative maintenance steps exponentially extends equipment's usable lifespan. But negligently abusing gear virtually ensures premature failure when reliability matters most!

A bit of care goes a long way. Be proactive in checking fittings, cleaning contacts, backing up data, and resetting systems to keep radios humming for years of service rather than mourning after a single season of abuse finally pushes components past failure.

10.3 Troubleshooting Your Baofeng Radio

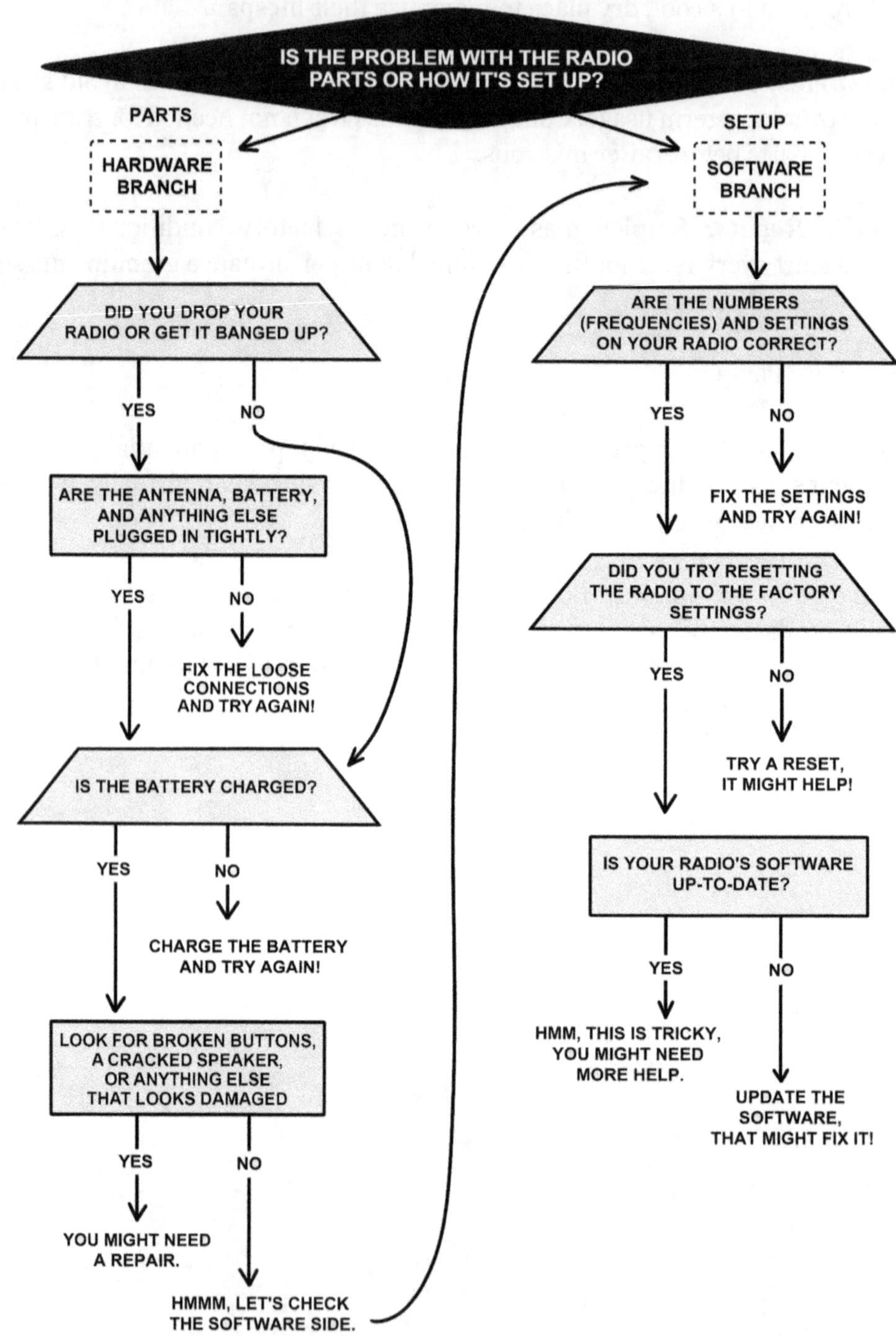

10.4 Extending Radio Lifespan

Quality Baofeng radios serve for years, even under harsh conditions, if properly maintained. Let's explore next-level tips to extend useful lifespan further:

Use Case Customization: Avoid needless over-speccing! Purchase models with ideal feature sets for YOUR requirements, minimizing unnecessary complexity risks. Unlike the most expensive triband tactical models, simple UHF units excel for many simple communication tasks. Right-size!

Weather Protection: Water, dust, and shock damage cut service life drastically. Invest in quality cases, weatherproofing accessories, and impact-absorbing mounts early to offset exposure risks in your operating environment. Don't gamble on gear survival - counter threats!

Power Options: Carry multiple charged battery packs when active use exceeds individual pack capacities. Rotate fresh cells instead of constantly recharging a single pack to maximize total charging cycles before degradation. And protect that charging port!

Careful Storage: Keep unused units away from temperature extremes and wet environments. Remove cells from radios during extended downtime to maintain a battery charge of around 50% to avoid deep discharge damage. Exercise units every few months to prevent component seizing.

Pre-Mission Inspections: Before every primary use, thoroughly test programming, controls, connectivity, and reception quality. Verify no degradation needs to be addressed from the last use. Nip issues proactively!

Use OEM Parts: Stick with original manufacturer-approved replacement batteries, antennas, and accessories for assured compatibility and quality. Avoid third-party knockoffs, often skimping on materials and tolerances, introducing failure risks and radio damage.

Lifetime Value vs Cost

Consider total years of usable service when assessing radio value, not just initial purchase costs.

A radio costing $80 but reliably serving just 1-2 years before replacement needs may ultimately cost more long-term than a quality $140 radio easily exceeding 5+ years of active life through careful maintenance.

Pay more upfront for trusted brands and reputable sellers to maximize returns on your investment and minimize replacement hassles down the road. Take the long view crafting robust radio solutions!

In this chapter, we've provided in-depth troubleshooting and preventative maintenance. Being comfortable diagnosing and resolving common radio issues while knowing best care practices keeps your team communicating reliably.

Remember to periodically revisit maintenance steps even if the systems are operating fine. Staying vigilant ensures you don't take continued performance for granted. Then, when problems inevitably arise, you'll be equipped to efficiently return the radio to service and complete the critical missions relying on it!

With these self-sufficiency skills, you maximize usefulness and get the most from your radio purchase. Let me know if any topics need more explanation or if specific issues arise. But for now, operate with added confidence!

CHAPTER 11: EXTREME ENVIRONMENTS USE

Using radios in really harsh environments, such as wildfires, freezing Arctic weather, ocean storms, and earthquake zones, usually reveals the weakness of the radio design.

Fragile electronics fail much quicker when exposed to salt fog corrosion, bomb-like shock forces from drops, gagging smoke clogging ports, etc. Sensitive gears are also easily exposed on ships, mountains, disaster sites, etc.

However, picking the right rugged accessories and carefully managing radio use can still allow for surprisingly good performance even in some extreme situations, such as waterproof cases, signal booster antennas, and group call limits.

The key is smartly anticipating the radio's operation in extreme conditions. Then, choose sturdy enough devices and know how much you can rely on them in the field. Refrain from assuming electronics will be reliable nonstop when faced with Mother Nature's wrath.

This chapter discusses accessory options and usage tactics for extending two-way reach when demands become difficult. Let's maximize these radios smartly.

11.1 Configurations for Extreme Conditions

Let's start by looking at what environmental conditions your radios need to operate in. Weather, shocks/drops, dust, debris, and anything can damage the electronics.

Weather

Radios in hot, wet, or humid environments can damage them over time. Electronics prefer to avoid temperature extremes, heavy rain, or thick moisture for a long time.

Plan for your radio to face things like 100°F (40°C) heat waves, icy cold winter nights, huge raindrops in forests, and soggy humidity near oceans. That covers most global weather risks.

Assume the radio will get splashed, soaked, or even dropped entirely in water at some point. Don't expect it to stay nice and dry.

The critical point is understanding how much temperature, water, and humidity exposure your location could experience. Then, choose radio models rated to handle rain, heat, and moisture over long periods.

Check official specs for operating temps, water resistance ratings, humidity tolerance, etc. Pick radios, proving they can survive extreme testing even for hours when soaked or baking under a hot sun. It'll be worth it out there when the weather inevitably strikes.

Shock & Vibration

Your radio will take some beatings in the rough world. Consider how much hitting, dropping, and crashing it may experience in the wilderness or disaster sites.

Exceptional military shock rating standards help compare how much bump forces different radios can handle before breaking internally. It's measured by G-force amount, like the intense snap of quickly turning fighter jets. More G's mean surviving more considerable sudden impacts if dropped or smashed.

Any radio you depend on should have a MIL-STD-810 shock rating level in product details. This specifies tested toughness, like sustaining 60+ G inertia hits absorbed without damage. Your radio needs to be able to take many hard falls.

The overall point is to pick radios built to handle significant vibrations, shaking forces, and sudden impacts. Ensure their published shock/toughness rating meets your hazardous needs. Read the specs to prevent early radio failures after hard hits during your missions. Sturdy stuff lasts the longest in the field.

Dust & Debris

It's not just weather that beats up radios—little specks like dust, dirt, sand, and ashes can also sneak inside and cause significant problems.

When large quantities of tiny grit are crammed into ports, it can jam buttons or kill mic sound. Bits rubbing on circuit boards scratch up essential paths. Food crumbs in sockets stop chargers from working.

The goal is to keep every possible speck out that could damage delicate electronics. Check protection ratings like "IP55," which indicate that the radio is fully sealed against dust entering anywhere.

Look for rubber gaskets around battery areas and waterproof flaps covering the jacks you plug into. Any tiny exposed cracks or slots allowing the smallest particles in will lead to issues when outdoors in the thick dust or sand for weeks.

So be sure to pick super sealed-up tight radios as much as possible for your gritty mission locations. That keeps all electronics safely shielded from problematic dirt/dust/sand ruining gear from the inside.

Post-Disaster Contamination

After terrible events like floods, eruptions, massive fires, and even war attacks, the air and ground become filled with nasty filth. Mud residues, choking ash clouds, and oily smoke residues contaminate areas.

Even the best-protected radios won't hold up long when coated by corrosive chemicals, clogging soot, and conductive crud getting smeared into ports by hands during urgent efforts coordinating help.

So intelligent leaders assume radio performance will degrade and fail quicker than expected when used in challenging post-disaster environments filled with residue contaminants.

If conditions allow, carefully keep gear off whenever possible to remove grimy build-up and acknowledge that hardware lifetime drops sharply depending on contamination exposures. Backup radios and repair parts become vital considering shortened lifespans.

The main takeaway is that if responding to crises in contaminated zones like wildfire ashes or coastal flooding sediments, anticipate replacements needing accelerated rotation as filth gradually turns off even rugged electronics.

Here's an easy toughness comparison table for Baofeng radio models:

Baofeng Model	Water Resistance	Shock Resistance	Dust Resistance	Overall Toughness Rating	Notes
UV-5R	Low	Low	Moderate	★★★☆☆	Budget-focused, not for extreme use
BF-F8HP	Moderate	Moderate	High	★★★★☆	Affordable, splash-resistant
UV-82HP	Moderate	High	High	★★★★☆	High-power output, water-resistant
UV-9R Plus	High	High	High	★★★★★	Best overall toughness in the Baofeng line

a- Preparing Ahead

Once anticipating likely exposure threats, purposefully select accessories and configurations countering potential damage/performance degradation:

Rugged Cases

You can buy hardened protective cases custom-fitted to many radio models that surround the electronics. These rugged shells act like armor-securing gear inside.

Ensure any case you get is rated waterproof and shockproof to depths/impacts matching field use. Inspect possible open places where water can enter the radio closely.

Companies advertise glorious claims about product toughness and resilience. But before totally trusting marketing hype, double-check user reviews from disaster teams who have used and tested the radio to see what they say about it.

The proper aftermarket radio case barricades electronics from damaging conditions long enough to accomplish the mission. Just confirm verifiable effectiveness data first-hand before assuming labeled "protection" unconditionally saves radio longevity when environmental factors assault functionality. Independent verification counts!

Weatherproofing

Protective sealants or films like Liquipel provide water-resistant barriers for exposed radio ports and circuit boards that are vulnerable to moisture damage. These coatings cause water to bead up and roll off rather than penetrate vulnerable interior electronics. Also, for easier fixing, you can find rubber stoppers or port caps sold that tightly fit into transmitter jack openings, battery spaces, etc., on most common radio types. This keeps some moisture out. Just inspect if total water "proofing" is reliable in the long term for extreme conditions that are expected sometimes.

When installing weatherproofing treatments, do not seal over speaker openings or microphone holes that must remain permeable for proper audio functionality. Only hydrophobic layers are applied to non-critical outer housing surfaces.

Reasonable weatherproofing applications shield sensitive radio componentry from rainfall and other wet threats. However, ventilation space for heat dissipation and unobstructed audio ports are essential for operational performance longevity in harsh outdoor conditions.

External Controls

You need models with big, easy-to-use buttons when using radios in freezing weather or wearing bulky protective gear. Tiny controls become impossible to press precisely when wearing thick gloves.

Getting radios advertising "glove-friendly" interfaces ensures easy operation without the finger-level finesse necessary. Even pressing transmit power buttons should activate reliably even when losing tactile sensation from numb, cold hands over long shifts.

Emergency teams in risky areas depend on quickly accessing critical functions with restricted vision/mobility by wearing fully encapsulated suits. Evaluating external radio controls using realistic mission constraints prevents field communication impediments when responsiveness matters most. I don't find out if the buttons need bare fingertips mid-crisis.

High Visibility Controls

Using radios when visibility is terrible—like smokey fires, dense dust clouds, or torrential rain limiting vision—becomes extremely difficult if crucial buttons and screens are more apparent.

Tiny, crowded Baofeng displays with cluttered controls make it nearly impossible to read quickly when visibility drops. Getting models with backlit keys, large primary buttons, and high contrast markings helps operators access critical functions even in obscured environments.

Emergency transmits tabs need noticeable alert colors and oversized grip surfaces that facilitate urgent activation even when blinded from the primary senses. Steady glow striping along device edges aids in quick retrieval when fumbling around in the darkness.

Maximize vital radio controls with gripping cues sensed beyond just vision alone. Requiring precise eyeball placement on confusing interfaces mid-crisis causes significant lag when instant reach matters most. Hectic conditions demand intuitive equipment!

Handling Temperature Extremes

Radios need to Turn ON and run when ambient temps around them rise or drop to crazy hot/cold levels - not just lovely spring afternoons.

Double-check if the official operating temperature ranges in the product specs meet realistic peaks your location could see annually. Don't assume advertised values reflect survival extremes; request clarification if unsure.

Many low-cost radios only function at 32°F (0°C), freezing before electronics start glitching from stiffer solid-state components. But the frigid Arctic and alpine use requires supporting start-up and stable transmission closer to -4°F (-20°C) ambient or lower before wind chill!

When a radio says it handles temperatures up to 105°F (40°C), read closer to see if those are brief spikes. Cheaper models survive intermittent highs, but sustained heat still kills circuits.

Military-grade radios must run 48 hours at even hotter temps near 140°F (60°C)! The difference is whether 105°F spec only allows 30 minutes before overheating fails components.

Also, some radio ads boast excellent temperature ranges on their gear handles. But the gear reveals weaker real-world limits once it has been out of comfortable room environments for a long time.

Companies try to impress buyers with cool buzzwords like "all-weather capability." Only trust sales hype about operating in extremely hot or cold conditions by double-checking first!

Independently verify promises, ensuring transient, cozy operating temps equate to HOURS functioning nonstop once ambient extremes hit the radio. Don't gamble group coordination on marginal resilience absent evidence.

When picking mission-critical radios, independently validate marketed durability claims against worst-case scenarios you could face. Don't risk dangerous capability gaps once environmental factors severely impact electronics for sustained periods. Research specs carefully using scientific sources rather than believing bold sales claims alone, promising the world arrogantly. Vet gear rigorously—lives depend on it.

Preventing Internal Condensation

Sudden temperature swings from hot to cold allow moist air trapped inside radios to "fog up" as conditions change. It's like eyeglasses fogging, going from freezing outside to warm indoors.

When temperature shifts occur rapidly, tiny water droplets form on electronics. This moisture corrosion can damage radios over time.

Carefully drilling tiny drainage holes in the radio case equalizes internal/external air pressure differences, causing water to be collected. This stabilizes the humidity and temp inside to match the ambient environment.

Be careful not to drill too many large holes, as this could compromise the radio case. Also, avoid putting holes anywhere critical components could be impacted by liquid or debris leaking through! Precision matters here.

The goal is to safely relieve internal pressure variance zones that cause condensation on crucial circuits when environmental factors swing wildly across freezing points.

b- Use Radios Very Carefully

Even with excellent rugged radios, be extremely careful about using them in harsh environments. Don't take access for granted.

Things like battery charging, signal-boosting towers, and quick replacement parts won't exist in the wilderness or disaster sites. Excessive use of delicate electronics can result in abuse, so know when to use your radio.

Avoid battery drain by minimizing unneeded chatter that shortens talk time later. Stop interference by spacing out updates instead of everyone talking over each other chaotically. Avoid damage by holding the radios carefully instead of casually holding them unprotected.

Leaders must guide teams to resist constant radio contact dependencies in remote field sites. Set expectations for periodic conservative check-ins conserving battery/airtime. Don't let folks think 24/7 reachability stays guaranteed despite a lack of support systems!

Electronics fail catastrophically once infrastructure crumbles. Avoid potential isolation by carefully scheduling radio usage for only urgent coordination. Keep usage minimal and not to be used for non-critical situations. Handle gear with tremendous care.

11.2 Communication Strategies

Radios are fragile without infrastructure. When away from power grids and repair shops, even "rugged" radios become serious failure points for teams if mishandled. Their limited battery and hardware lifespans require careful use; therefore, ensure that you put all necessary measures in place to help you communicate better on-site.

Coordinate Communication Schedules

Groups must agree on radio check-in schedules to avoid draining power with casual chatter. Set specific timed windows like "10 and 40 past the hour" for all teams to briefly update statuses.

Concentrate listening during these exact minutes and resist unnecessary chatter distracting actual emergency calls. Keep to crucial contact without constant casual back-and-forth.

This disciplined rhythm focuses on critical update times while preserving battery capacity across all distributed radios. Stick to urgent coordination only outside designated sync-up intervals.

Minimize Transmissions

When everyone is stuck together in close quarters like a crowded shelter, limit radio usage to prevent painful audio feedback (loud squeals) as signals echo back.

Appoint one person to step outside using the exterior antenna to coordinate outside communications. This allows amplified broadcasts without deafening folks inside from close feedback interference.

Teams should collect all needed information internally before a designated member summarizes key details externally to other groups. Refrain from cluttering channels with scattered overlapping calls.

Leaders must enforce discipline even during emotional times when too many passionate people shout chaotically. Clear communication requires controlled coordination despite tensions. Prioritize clarity.

The key points are to consolidate internally first, then assign specific members to relay outward selectively. Prevent feedback by isolating speakers. Keep messaging tight instead of multiple mixed shouts cluttering up limited channels. Structure enables progress.

Have Backup Communication Plans

Things go wrong in the field, so intelligent teams prepare backup options, allowing radios with dead batteries or damage to call for help if separated.

Come up with contingency plans like:

- Memorizing emergency frequencies.
- Distress signal patterns.
- Quick response code words.
- Secondary rally points.

Share these backups so all members can reach assistance if primary radios fail unexpectedly.

We only partially depend on direct contact in highly remote areas. Having redundancies and knowing protocols if hardware suddenly dies allows reconnecting or escalating emergencies, saving isolated lives.

So, put together alternate channels beyond primary radio frequencies and drill backup details extensively with the team. Crucial contingency preparation prevents catastrophic loss of coordination when operating off-grid if equipment gives out.

11.3 Vital Accessories

We can prep personnel and plans, but sufficient equipment ultimately enables execution. Let's examine essential accessories for further shielding radio hardware itself when operating off-grid under demanding conditions:

Dry Boxes: Certified waterproof cases like those from Pelican protect gear from driving rain, river crossings, and sudden immersion risks. Ensure a snug, pressured fit with desiccant packs for added moisture protection.

Hard Shells: The impact-resistant polycarbonate plastic or TPU polymer cover absorbs accidental drops and shocks that easily shatter naked electronics. Bonus points for snap-on clarity, allowing interface access.

Shoulder Straps: Rather than casually stowing loosely in packs, strap radios securely to harness anchor points, keeping units handy and protected against violent dislodging even through sustained bouncing motions.

Spare Batteries: Prevent dead air threats by packing backup cells for quick hot swapping internally or through external recharging cases. Standardize additional power cells across your entire team's radio fleet, allowing compatibility with universal spares.

Signal Boosters: Portable mini signal amplifiers like the popular MMDVM boost reception for small transmitters when teams spread across distances exceeding hardware capability in remote areas. Extend viable range.

Custom Cables: Fabricate and pack short jumper cables bridging damaged ports externally should components like microphones, antennas, and headsets tear away and must be bypassed until total radio repair is done later. Strive for universal compatibility between all makes/models, equipping members.

Having rugged gear and setting good rules that prevent problems before they happen makes all the difference when things get crazy in tough spots.

Teams that waste time scrambling to hack solutions mid-emergency often regret skipping preparations earlier. Don't cut corners in advance only to regret it later.

Careful planning for worst-case scenarios, backup policies, and survival accessories removed delays and doubts confronting extreme chaos. Reasonable readiness tragically divides resilient groups from struggling victims.

While nothing promises 100% certainty, reasonable contingency prep steers teams safely through most storms.

Remember, smooth operations start with leadership building comprehensive preparations long before sudden events challenge response capabilities. Invest wisely in systemic protections enabling stability when catastrophe strikes!

CHAPTER 12: INTEGRATION WITH MODERN TECHNOLOGIES

Baofeng walkie-talkies work great for essential talking between people. But these little radios suddenly become more powerful when you hook them to modern gadgets like forest wildlife cameras!

Let's check out some cool next-level gear we can pair up to turn basic Baofengs into ultimate radio data hotspots. By tapping into technologies like the "Internet of Things" (IoT), we'll beam more info faster to headquarters from many places, even where the internet can't reach.

When you integrate the flexibility of affordable radio equipment with high-tech sensing for automation networks, ample opportunities open up, extending our communications capabilities tremendously over long distances. In future calls, let's explore some cutting-edge pairings.

12.1 Baofeng Radio with IoT Sensors

"IoT" describes a massive technological shift - ordinary objects now sense environments and share data inputs through the internet, allowing centralized monitoring and control.

This means regular physical things like home appliances, infrastructure components, transport vehicles, or environmental sensors wake up with computing brains.

Interlinking these intelligent endpoints creates expansive sensor-based networks that generate/consume huge dataflows. IoT revolves around distributed smart objects sharing contextual telemetry that is customizable according to needs.

IoT Sensor Opportunities

So, what does the IoT explosion mean for traditional radios? New roles are interfacing abundant embedded technologies.

Consider fresh pairing possibilities:

- ◆ Trail camera image bursts.
- ◆ Pipeline valve monitors.
- ◆ Waterway flood gauges.
- ◆ Firewatch thermal scans.
- ◆ Air quality sensors.
- ◆ Bridge strain detectors.
- ◆ Storage tank monitors.
- ◆ Perimeter access controls.

These present just a small sample of cases where radio-relayed data channels prove vital in capturing sensor insights from distributed monitoring nodes.

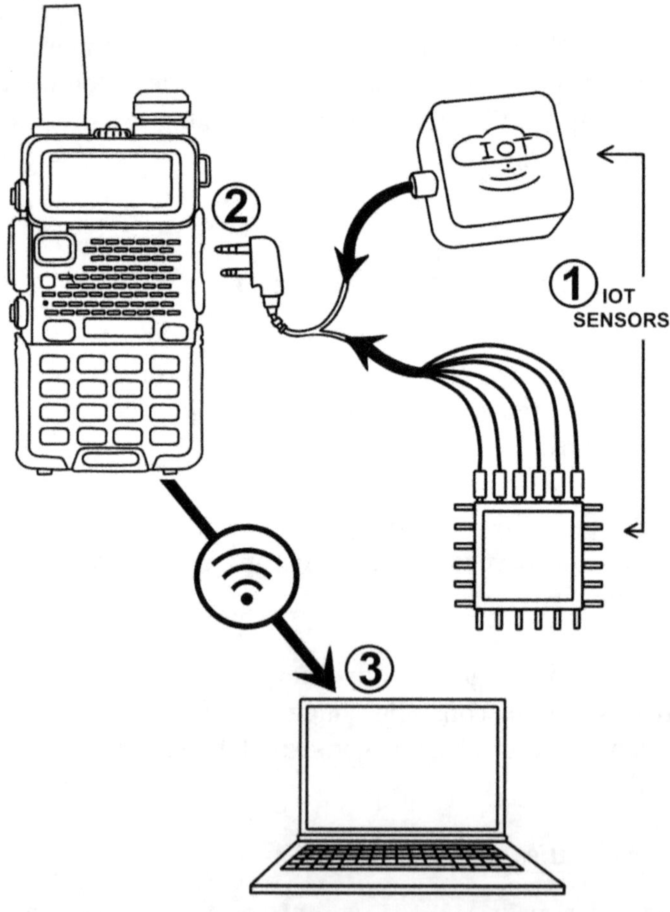

CONNECTING RADIO TO IOT GADGETS

Baofeng Radios Act As Data Bridges

Here's where affordable Baofeng walkie-talkies truly shine - gathering remote sensor data then beaming it wirelessly so it doesn't get stranded.

The key is converting all that sensor digital data into audio tone patterns like old dial-up modems. Custom microcontroller code helps package telemetry bits into squelch signals that Baofeng speakers/mics can transmit as everyday analog voice data riding FM radio waves.

While not lightning fast, these modulated frequency patterns carry sensor readout information surprisingly far over radio channels matched to this data bandwidth capacity.

The flexibility lets us relay and monitor device metrics from very remote field locations lacking traditional internet access over wider areas than is expected. With creative audio packaging on one end and proper radio reception at headquarters, Baofengs bridges the data gaps.

Use Cases Abound

There are endless ways companies can now easily connect monitors wired up with IoT smarts to transmit data over underused Baofeng radio channels.

For example, utilities could overlay sensors across miles of remote pipelines. Data is relayed wirelessly to headquarters over the existing radio network instead of needing costly cellular links.

Also, scientists studying wildlife in preserved forests can tap motion-activated cameras. Images relay immediately instead of risky monthly pickup, extracting location data revealing animal habits. Thanks to radio flexibility, no new infrastructure is required.

The key is creatively using available radio equipment in totally new ways to support a flood of incoming sensor data. With the proper tuning, affordable Baofengs become decent at catching remote monitor transmissions. They get upgraded into excellent little machine data rescue networks!

This taps potential that's been there all along using global radio assets. But nobody realized traditional FM voice walkie-talkies could provide such great sensor data bridges until IoT platforms took off, needing cost-effective transport mediums. That's pretty innovative thinking, I would say.

12.2 Smart Cities Applications

Beyond remote stuff, Baofeng walkies also integrate nicely into high-tech city infrastructure. Communities gain radical responsiveness and efficiency by linking things like streetlights, alarms, and transport together using shared data.

The idea is to optimize electricity, traffic, and safety resources by enabling all equipment to self-monitor conditions while interconnected. This allows super-fast coordinated reactions as sensors pool data about precisely what's happening everywhere.

Picture extra buses dispatched precisely as passenger demand spikes or traffic signals adjusting green time automatically to favor current road congestion. Resources get deployed on-demand, guided by a shared city data web. Become a brilliant instant-response metropolis.

But knitting so many data nodes across vast districts demands reliable equipment, bridging the gaps when networks inevitably get interrupted. Once again, this is where adaptable Baofeng radios became very useful.

Handheld walkies penetrate infrastructure dead zones with the proper setup, capturing readout bursts from local controllers. Then, data payloads are shot upstream into master analytics engines via creative radio transports pieced from existing voice equipment in a pinch.

Rather than total visibility loss when primary grids go down, Baofengs ensure graceful degradation, managing critical telemetry streams despite inevitable outages. Hardened radio flexibility preserves safety and responsiveness, cementing smart cities at scale.

Radios Step Up When Technology Fails

Imagine car systems detecting engine failures or dangerous driving, then auto-reporting issues through entertainment radios. Drivers get alerted about risks even before breakdowns through constant sensor monitoring.

Or take bridges scanned by vibration detectors, which spot tiny cracks early.

When other communication failures happen, this feeds constant structural monitoring and pipe telemetry data through secondary radio networks to prevent disasters.

Even if major smart city grids go offline due to cyber-attacks, interlinking backup radio systems temporarily keep some flow tunneling through multiplied pathways. It is not fast, but it keeps trickles of intelligence dripping to controllers, preventing total city-wide blinding during regional outages. Hardened hardware gracefully bridges the gaps.

Modern city efficiency depends increasingly on nonstop data streams interfacing equipment and response teams citywide. Embedded sensor readings drive real-time automation. So, when cellular networks drop, nimble radio systems swap as flexible data conduits, preventing notification halts. Signals may slow but won't entirely halt thanks to radio versatility plugging communication holes on-demand when disruption strikes regional centers. Graceful fallback options enable sustainable expansion, scaling intelligence broader despite inevitable tech failures.

In this chapter, we looked at some neat ways basic Baofeng walkie-talkies could pair up with high-tech sensors or city systems by relaying data creatively over regular radio waves.

With the proper tuning, these affordable radios act like resilient bridges, filling in gaps when regular internet or cellular networks get disrupted across regions. Keeping some flow of sensor data prevents total blinding of control rooms.

The point is that radio hardware flexibility helps architects design brilliant, responsive future cities, even with the risk of high-tech software or hardware failing occasionally—like a backup nervous system!

So, when planning modern intelligent metropolises, consider the versatility and durability of radio communications' survivability. This ensures sustainable expansion despite inevitable long-term outages. Harness in innovative ways.

CHAPTER 13: CREATIVE AND UNCONVENTIONAL USES

Radios like Baofengs do way more than just letting people talk back and forth. If we get creative, these little devices can do pretty neat tricks that the makers never imagined.

13.1 Innovative Applications

Let's look at inventive ways to use these affordable walkie-talkies in roles beyond their regular jobs. Once you start trying them out, the options just start opening up.

Beaming Signals Farther

Suppose emergency workers must stay in touch with headquarters by trekking deep in the wilderness, totally out of normal radio range. In that case, you multiply the transmission distance.

You can achieve this by combining high-gain antennas and maxing out signal amplifiers. We can blast radio waves up where they bounce off atmospheric bubbles in space, massively multiplying how far transmissions travel.

These cheap handheld walkie-talkies can extend coverage by many miles without needing expensive relay towers or equipment. This is perfect for remote teams sending health sensor data and requesting custom emergency gear be shipped to them, even in super rugged terrain.

We can even bolt mini radio repeaters onto high-altitude balloons or weather kites floating miles up to capture these bounced signals, strengthening them even more. Now, sensors on drifty gear connect solidly to distant receivers, resiliently routing data streams over the curved horizon as floating payloads meander into view.

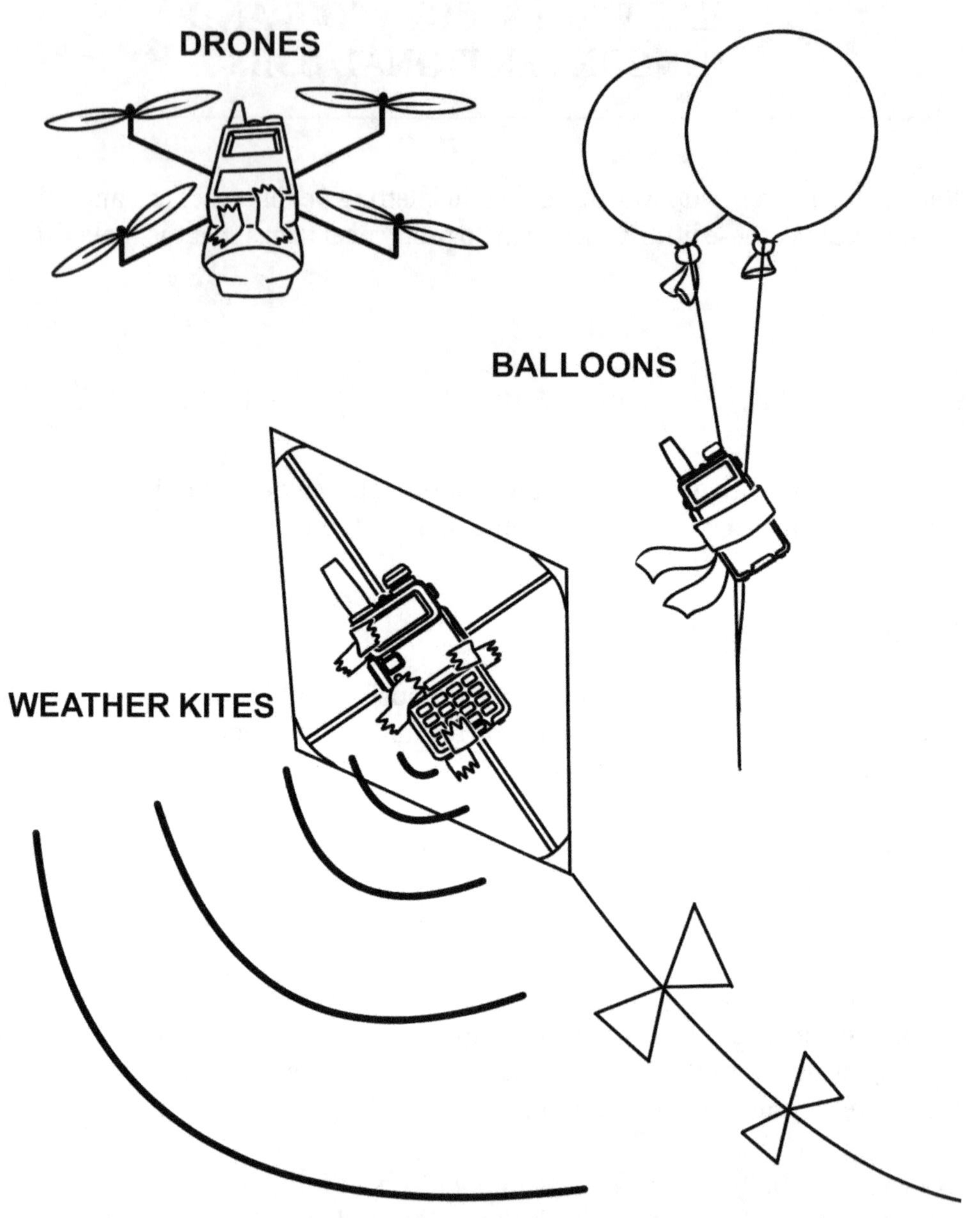

BEAMING SIGNALS FARTHER

Bridging Across Radio Systems

Disaster scenarios pose a tricky challenge. Emergency crews arrive on the scene utilizing designated radio frequencies but can't coordinate with local police departments because of incompatible signal protocols. This communication barrier severely hinders multi-agency effectiveness in collectively stabilizing chaotic emergencies. Radio differences shouldn't cost lives when rapid unity and explicit cross-team coordination prove vital.

Luckily, we can leverage a particular software-defined "cognitive radio" mode by hacking the Baofeng to actively switch between multiple frequency bands smoothly in real time.

This on-the-fly bridging across public service, business, and amateur radio spectrum blocks barriers posed by equipment disputes or protocol differences hindering multi-agency coordination.

When organization politics or bureaucracy threatens emergency effectiveness, creative Baofeng hackery shoulders tactical unity burdens itself, transpiring communication channels atop incompatible airwave infrastructure. Talk about leading the way and promoting common ground when it matters most.

Keeping Messages Secret

If we want to hide team chatter from folks scanning public frequencies, we can mimic military encryption by rapid "frequency hopping".

Custom hacks make Baofengs jump between dozens of bands algorithmically in a readable yet random-looking sequence known only to squad members. Signals fade noiselessly for outsiders, sweeping channels unable to crack shifting patterns quickly. Our messages hide in plain sight.

The benefit of encryption is lighter computing demand from basic radios, leveraging agile spectral jumps that intelligently sidestep attention that does not belong to missions. Analysts can't differentiate signal fading from background static with communications distributions dispersed probabilistically. Slight signal uncertainty gives us an advantage enough for operational needs.

So, with the right hackery, handheld walkie-talkies gain military-grade transmission secrecy, perfect for teams working on sensitive events below mass attention.

Getting Creative with Audio FX

Crafty hackers can tap Baofeng ports connecting analog audio effect chains, deliberately distorting clarity before transmitting again—like guitar pedals, amplifiers, etc.

That results in a deliberate, overdriven, raspy sound that becomes like a quirky signature for team members. The uniqueness also complicates casual listening from outsiders.

This action homebrew flair marks insider radio chatter apart with surprisingly difficult coherence for standard receivers to decode.

Given some electronics courage, creatively exercising technical specs, and experimentally tinkering with radio chains, the possibilities stay wide open. Of course, regional transmission regulations should always be respected.

13.2 User Customizations and Hacks

Beyond these hacks, all kinds of users upgrade Baofeng radios with handy personalized modifications improving regular real-world use:

Improving Grip and Handling

The smooth plastic radio bodies slip easily from wet hands when moving fast. To improve grip security, wrap 3D-printed textured handles around the chassis.

We can also glue on rubbery pickup grips, anchor custom wrist straps, and even add microphone windscreens to prevent debris from striking the delicate internal mics. This makes a big usability difference when using radios outdoors in demanding situations.

The goal is to improve ergonomics and protect user experience by improving rugged vulnerability points like mics or screen lenses.

Sealing Vulnerable Ports

Outdoor teams should protect exposed radio ports from weather threats like blowing sand, rain, and snow.

Snug rubber caps stretch over jacks and can secure them from dust and moisture when accessories are unplugged without entirely blocking slots. Retention cords also prevent dropped mics.

You can use dummy accessory plugs to seal unused ports from particle side-attacks. Minor speed bumps can twist open access flaps, but securing delicate interior electronics from contamination is worth securing.

Avoid direct exposure on cables/jacks, risking corrosion or debris fouling and plugging things in later. Defend ports proactively.

Improving Visibility and Speed

Little Baofeng buttons can be tricky to manage quickly with gloves on or when vision is limited.

We can install external oversized silicone buttons, tactile cues, and hand grooves to speed complex menu actions into single-press macro shortcuts. This muscle memory accelerates emergency access even in the dark while enhancing grip stability expected for sustained outdoor use.

Don't tolerate preventable UI constraints—creatively customize interfaces to serve unique use case needs for optimal speed, efficiency, and error resistance. Interfaces can be tailored to resemble operators' extensions, thoughtfully customized for critical visibility and easy access.

Get creative improving capability interfaces directly rather than tolerating incremental annoyances needlessly limiting workflows. Commercial packaging always needs to address actual field usability, so proactive customization boosts specialized ergonomics to deliver optimal use when reliability matters most to operational success.

As we can see from this chapter, the Baofeng radio can be customized to give something far beyond just talking. Their flexible hardware supports all kinds of creative upgrades if we reimagine applications broadly.

These include environmental sensors, system gateways, signal range extenders, and audio distortion tricks. But we must courageously approach radio gear, embracing unique edge-case thinking that defies simplistic sticker feature lists.

CHAPTER 14: REGULATIONS AND RESPONSIBLE USE

While tapping these radios' flexibility feels liberating, we still must follow some essential rules and use them conscientiously...

14.1 Licensing and Regulations

Recreational use of compact walkie-talkies is uncomplicated. However, transmitting radio signals falls under government oversight in managing airwave infrastructure. We'll look at some key concepts securing legal operations.

International/federal telecom authorities administer spectrum licensing to ensure tidy frequency allocation across public/private services. Regional bodies govern local standard compliance. Memorize correct overseers for your location.

License Requirements Vary

Personal two-way radios typically utilize unlicensed bands if transmitting below broadcast power thresholds, simplifying small-group adoption. However, expanding capabilities may necessitate more advanced operator or equipment credentials.

For example, GMRS operation in the 462-467 MHz range merely requires a no-fee license covering family members. Yet accessing extensive amateur bands requires passing rigorous tech examinations, earning call signs licensing your control, and granting expanded spectrum privileges.

Learn strictly which bands and radio capabilities are permitted freely and which require proper local licensing. There should be no excuses for skirting rules.

Stay Legal with Best Practices

While enforcement varies locationally, ignorance won't excuse violations once caught. Safeguard rights responsibly:

- ◆ Transmit only on designated frequencies.

- ◆ Reduce power/duty cycle if needed.
- ◆ Never interfere with prioritized services.
- ◆ Avoid signal boosting near airports.
- ◆ Report any apparent policy conflicts for clarity.

If you need more clarification about radio rules, it's wise to equip yourself with the necessary knowledge by double-checking with qualified experts rather than guessing or assuming incorrectly. Reach out to telecom authorities in your area for guidance about legal compliance rather than risking illegal activity. It's okay to confirm specifics if complex regulations seem confusing to you.

So, all that is being said here is that personal radio device rules vary globally and can feel inconsistent. So, to be wise with your experiments, always bookmark helpful legislative references you can recheck easily. Save links and numbers for informed oversight contacts in your area whom you can set up informal video consultations with before ever rigging or keying up new equipment tests that end up accidentally illegal due to assumption mistakes or complex misinterpretations. Being super careful upfront and asking plenty of questions prevents painful penalties if problems ever get reported needing follow-up.

14.2 Safe and Responsible Use

Beyond getting broadcasting licenses, we need to be extra careful about potential signal disruptions nowadays since many devices use crowded airwaves. Our radio activities can't carelessly stomp on another bandwidth—we have to share the spectrum calmly and cooperatively.

Listen Before Transmitting

Always briefly scan through the dial before transmitting to double-check that your local channel is clear and inactive. Even if you legally have rights to broadcast there, barging onto a frequency already in use generates chaos and headaches for users trying to chat orderly. Avoid blasting onto existing conversations to be polite! Give other radio operators calm space.

The critical etiquette is to actively look first before transmitting on bands. By checking first, you can only assume a frequency is precise or unused from a long distance. This will save people listening headaches!

Use the Lowest Power Possible

The key idea when transmitting is to use only the minimum amount of signal power necessary, for the shortest duration necessary, to stay connected to your intended people without issues. You don't want to blast out extra high-powered signals that "bleed" onto and crowd up neighboring channels needlessly. Keep things minimal but crisp to be efficient and polite to others using nearby bandwidth.

Coordinate Channel Assignments

When operating multiple groups, intentionally assigning discrete radio frequencies to each team right from the start prevents bandwidth waste by having everyone chat chaotically on top of each other. Reserve separate clear lines for focused coordination between individual groups.

And periodically change the specific channels teams rotate rather than permanently lingering on fixed pairs. This avoids interference patterns that can crop up when equipment stays tuned to unchanging spots on the dial for an extra long time when unchanged. Mix things up now and then!

That means you must be very intentional about coordinating bandwidth across multiple groups operating. Refrain from letting things get chatty, crossing lines, or idling channels extensively. Tight etiquette keeps the radio flowing smoothly.

Report Issues Politely

If odd interference starts disrupting your team channels, calmly reach out to local telecom managers about better containing the situation without anger or accusations. Politely ask for advice and assistance investigating the root signal causing problems. Provide any recordings you captured that could help pinpoint sources accurately.

The motto should be friendly progress over finger-pointing as technologies increasingly crowd-limited airwaves. There will be growing pains, but radio systems can coexist cooperatively long-term through respectful self-regulation and communication.

That is to say that by responsibly planning usage, being willing to shift frequencies to avoid persistent interference, and collectively reporting issues without blame, modern radio flexibility stays viable if we smartly balance sustainable shared infrastructure needs.

Well, we've covered a lot about radio licensing and legal usage. Understandably, some rules and policy variations between areas can initially be a little complex for casual users. However, starting with some essential self-guided learning goes a long way to avoiding problems or accidental interference.

Hopefully, some of our shared ideas can give you more confidence going forward with legally customizing channels, ranges, etc., as long as the equipment stays compliant. Just remember to double-check specifics with experts anytime capabilities start bordering advanced domains. Asking questions is always an intelligent way to handle things.

We covered the critical etiquette points about minimizing disruption - scanning first, lowering power, shifting frequencies cooperatively, and reporting issues politely when they rarely occur. Treat airwaves as precious shared roads, and we'll sustain access.

The motto to leave with is to progress collaboratively over finger-pointing if we encounter occasional growing pains. There's room for all of us reasonably.

Most importantly, have fun with the radios legally and safely. Call for assistance anytime questions arise, or experiments cross into more advanced territories. Fantastic local ham operators are happy to provide insights in most locations.

BONUS CHAPTER: ADDITIONAL RESOURCES AND COMMUNITY

In the previous chapters of this guide, we've covered a lot about Baofeng, but there are other aspects you can still explore about these radios. We'll talk about helpful materials, online communities, and skill-building resources to help you further with these radios if you want to level up from status later.

There is still a world of ham operators, emergency responders, and electronics repair wizards using advanced gear you could try your hands on. Once you've got the starter Baofeng techniques down solid, we'll explore some jump-off points for diving deeper. The radio adventure can keep going as far as you want to follow the frequencies.

15.1 Broaden Your Skills

Many new Baofeng radio users can gradually work toward getting their FCC amateur radio license. Passing the exam allows you to communicate more freely across wider bands, letting your messages travel much farther. Local ham radio users usually love guiding newcomers through licensure over casual weekly study groups that make learning easy-going with tested tips.

Attending big ham radio fests nearby brings everybody together to share their wireless tricks freely. At field day gatherings, you can kick back and try serious long-range equipment under the supervision of veterans who love to mentor about their crafty gear hacks. Their technical curiosity, voluntary teaching patience, and eclectic personal rig configurations make progress faster and more intriguing than wading solo through dense internet manuals you may find lifeless.

You can join regional ham associations online or in person; no question is too small. A friendly community awaits tapping into global networks. Just bring an open mindset and avoid assumptions to maximize modern radio purpose expansion possibilities creatively. Incredible journeys await with some commitment through abundant guidance.

15.2 Online Forums and Groups

Beyond static print guides, you can use dynamic radio operator communities for personalized advice. Here are some online knowledge sources.

Reddit Channels

Reddit subgroups focus on radio topics like HamRadio, BeginnerPrepper, and Baofeng. These discussion groups compile questions and advice about equipment troubleshooting or setup tours when radios are used in severe field deployments. Get amateur tips from fellow users to connect seamlessly and tackle spotty zone coverage or hardware problems. You can also get super helpful stuff from pro forums.

Facebook Radio Owner Groups

Facebook groups focus on radio models like certain Baofeng walkies or Cobra CB radios.

These groups let owners share advice on gear upgrades, custom add-ons, repairs, etc., centered on specific radio units. They become like a searchable knowledge base preserving tribal tricks perfecting particular radios through years of members sharing the latest modding techniques as hobbyists steadily bring up creative new uses.

Private Emergency Comms Groups

There are also closed private radio forums that help coordinate emergency responders logistically if public airwaves get too overwhelmed during disasters.

These secure contingency channels act as backup plans to organize critical relief efforts privately when open radio bands are too messy.

So emergency crews maintain operational security, recovering mission effectiveness after initial response stages, which inevitably grow chaotic publicly across civilian channels.

But for closed channel access, you must know the exclusive emergency radio groups and their reputations. Strict entry barriers intentionally screen participants, managing to uphold reliability. It's about trust when coordination carries life and death stakes!

Just be aware that private emergency channels exist discretely if all else fails, and the core situation spirals. Those hidden tactical frequencies shore up last-resort communication, able to cut through the chaos—when used judiciously.

15.3 Advanced Skill Building

Even expert masters remain lifelong students, constantly developing expertise over decades. We should never stop challenging our radio comfort zones. There are always new frequency arts to perfect.

Once you get good at basic stuff like swapping Baofengs batteries and plugging in antennas, you can move onto super-advanced telemetry and modulation mastery levels:

DIY Electronics Repairs

Imagine your radio fritzes out miles from the nearest store. You may want to Learn to diagnose issues, isolate internal damage, and even surgically solder-fix sensitive circuit boards yourself to resurrect dead gadgets.

With online repair video guides walking through equipment like Baofengs, you can carefully rework delicate electronics back to life by tracing printable schematics detailing exact board layouts and parts needing fixes.

It takes some 3 AM practice sessions first under the desk lamp to get your soldering precision ninja steady. But with enough dry runs and honest remote breakdowns, your manual repair skills become mission-saving legends, and you can securely restore finicky functionality from scraps in the field if needed!

Direction Finding

Here's an astonishing advanced skill: using antennas to track signal directions and locate the location of weird unknown transmissions popping up. It's like radar but for radio waves.

You can learn to manually plot the exact coordinates of sneaky interfering broadcasts messing with your team frequencies if they appear suddenly. Just point directional antennas around while listening for the signal getting louder as you zero in hotter on the target.

The techniques allow pinpointing any transmission source points precisely once you practice the triangulation methods enough. It is handy for figuring out where unexpected disruptions come from so you can fix the interference. With that, you've become an expert at detecting frequency.

Radio enforcement also uses it legally to find unlicensed broadcasters, for example. However, mastering radio enforcement as a hobby requires decent technical skills, such as controlling directional beam antennas precisely while interpreting signal clarity and waves on spectrum analyzers.

But if motivated, you can advance from passively listening to aggressively hunting down any area radio bands to uncover unidentified transmissions. This lets you lead investigations fixing interference. That is wisdom through practice right there.

So, there is something to gain from some of these references we have shared. The book materials, online groups, and skill recommendations hopefully open your mind to incredible possibilities if you ever feel like going deeper into trying things about radio.

There's much more waiting across the global radio hobby for every interest - from soldering circuit boards to hunting secret signals or building weather balloon rigs. If you choose to dive deeper later with new radio goals, consider this chapter your gateway to advanced radio worlds awaiting exploration.

Enjoy all the Baofeng fundamentals as a solid basis to chart your path. But whenever you're ready to level up your technical hobby skills or join online radio enthusiast communities to push equipment creatively further, you know where to flip for those frequencies next!

CONCLUSION - MASTERING YOUR BAOFENG RADIO JOURNEY

After so many pages helping you learn, we covered a lot together about using Baofeng radios—from turning them on for the first time to making them do advanced tasks.

It's like we started at the bottom of a tall mountain. I showed you step-by-step how to climb up. Now, let's pause and look back at the view from being way up here after that long hike upward. You can see more of Radio Land now than before when you were down at the trail start.

We'll review the radio basics I introduced in Chapter 1. Now that you know many more advanced things, it's easy to understand that beginner material.

So, take a minute to appreciate how far your radio skills have come. Then, use your excellent elevated view to find out what else you can learn next using these neat little Baofeng walkie-talkies.

I. Revisiting Key Goals

Let's go back and remember why we originally wanted to learn about these flexible little handheld radios:

A. Enabling Effective Personal Communications

Many readers probably wanted to learn about getting started with Baofeng radio for local communication—like field trips, exploring outdoors, youth group stuff, small businesses, or volunteer security patrols.

Even though they're very affordable, Baofeng radios deliver excellent basic two essential communication out of the box. Friends can chat short distances with the included pieces while balancing range and battery time.

We built a ton on top of that plain radio start by programming custom channels to match needs, creating private group codes, adding better hardware, and even getting the radios creatively to work with other tech gear.

Started as simple out-of-box walkie-talkies transformed into supporting excellent expanded uses on-demand - all while becoming super friendly for beginners to operate smoothly through practice.

B. Preparing Emergency Scenarios

Disasters can hit any time without warning - cutting off groups from central help when everyday communication tools like phones fail.

A backup radio setup that is good enough to allow teams to keep each other informed despite infrastructure breakdowns saves lives when racing to fix problems. Readers should now feel more ready to stay aware of what's happening in an emergency.

We prepared carefully for crisis scenarios using great Baofeng features like weather band reception, flashlight modes, pre-programmed safety alerts, and extra-powered batteries. Readers also learned innovative protocols to use the radios less often to make the batteries last longer.

Despite the chaos, communication separates teams that can overcome tough spots from stranded people waiting for delayed emergency help. Walking through emergency plans here aims to give you confidence when bad stuff goes down.

C. Building Radio Knowledge Foundations

Some readers initially became confused because Baofeng radios seemed like "walkie-talkie" gadgets. However, they quickly recognized all the technical details, making them work smoothly behind the scenes.

After reading through all the chapters in this guidebook, you will have a much deeper appreciation of important radio concepts—like how signals travel, choosing frequencies, adding antennas and accessories, programming settings, troubleshooting problems, and customizing features.

You started just wanting a basic walkie-talkie to chat locally with friends. But I hope you've discovered that affordable Baofeng radios are super versatile tools once you understand how to configure and enhance them in the right way.

There's a whole world of radio communications to keep exploring. But wrapping up this guide, you now have an excellent knowledge foundation to start putting core skills into practice.

We covered everything from turning them on to extending battery life, using them outdoors, planning for emergencies, and even linking to sensors and other tech gear. I explained Baofeng radios step by step.

If you're interested in future electronics projects or amateur radio licenses, you're equipped with simplified beginner operation competency and frameworks for grasping more advanced radio concepts. Feel empowered by what you've mastered so far.

II. Recap of Radio Concepts

Now, after reading a lot of information about what these radios can do and the rules for using them safely, you've gained a solid basic understanding of how these radios work rather than just seeing them as something to pick up and talk into.

Let's quickly review a few key ideas that we said would allow you to master your Baofeng radios better:

A. Radio Waves and Propagation

We learned that radio signals ride energy waves sent at specific frequencies based on the wave's length and speed. Changing the frequency changes how far it goes and what it passes through.

Readers now know radio spectrum segments (HF/VHF/UHF bands) have different traits that match their uses.

Hint - longer waves diffract farther, while shorter ones focus better directionally and penetrate buildings.

Tuning the antenna also improves efficiency by concentrating the signal instead of wasting it across antenna blind zones or mismatched gear. Careful alignment achieves good precision this way.

B. Frequencies, Channels and Privacy Codes

We went beyond vague "channels" to configuring send/receive frequency pairs saved into labeled memory banks for quick access later. Settings also assign squelch levels to filter out noise.

Programmable sub-codes like CTCSS/DCS numbers further divide who hears each channel. We use this like a gate to address specific listening groups fluidly.

Saving relevant frequencies into channel banks with contextual names surpasses hardware limits through the creative human organization without changing radio mechanics. Operational agility comes more from external intelligence than just what's inside the radios.

C. Accessories Improving Reception

Baofeng walkie-talkies become more helpful when hooked up with extra gear that tackles environmental problems. Smart add-ons empower the radios to handle challenging situations better.

Tough cases protect delicate electronics from weather, crashes, and dirt. Headsets clarify voices where background noise is high. Big antennas focus signals to reach farther and aim directionally. Backup batteries ensure extended run time when no power outlets exist remotely.

Clever Baofeng's use accepts adding supplemental hardware gradually to fill capability gaps in affordable portable pricing. We expand functions by blending thoughtful add-on upgrades into complete communication solutions.

III. Review of Models and Capabilities

While accessorizing bolsters capability, choosing ideal baseline models aligned with needs proves vital. Let's revisit popular Baofeng types and programming traits enabling customized implementations:

A. Leading Baofeng Models Overview

UV-5R: This feature-packed classic model offers dual VHF/UHF band flexibility through manual keypad input or robust computer programming support. It offers proven reliability at bargain pricing.

BF-F8HP: Enhanced receiver clarity plus boosted 8-watt transmission outage penetrating rugged terrain.

UV-82: Ruggedized, durable variant improving water/dust resistance for demanding outdoor reliability.

UV-B5: Streamlined UHF-only unit striking an affordable balance for no-frills communication needs, not requiring extensive frequency support.

UV-9R: Enhanced battery capacity and glove-friendly buttons ease outdoor access in harsh conditions.

Select ideal models that balance cost, complexity, and capabilities and support your exact activity needs. Avoid overspending on unused specifications!

B. Programming Features

Manually inputting frequencies and settings through a numeric keypad is tedious and needs descriptive context. Leveraging CHIRP software streamlines channel configuration drastically:

- ◆ Importing shared frequency spreadsheets across entire groups prevents individual typos.
- ◆ Organizing channel banks by contextual names/purposes enables intuitive recall.
- ◆ Backing up/restoring identical radio configuration profiles maintains consistency.
- ◆ Adjusting settings like transmit power levels, signaling formats, and selective scanning.
- ◆ Delegating repetitive manual input to computer assistance liberates groups in coordinating quickly.

C. Range Enhancement Customization

Beyond tweaking settings in software, physical hardware adjustments push range/clarity to surprising distances rivaling far more expensive commercial systems.

Careful antenna orientation, aftermarket gain upgrades, counterpoise ground plane wires, and height/line of sight exploitation squeeze every bit of reach possible from compact handheld Baofeng models.

Remember to consider what upgraded budget radios can do for long distances when artfully enhanced to match the outdoor areas you use. The right add-ons optimize their potential.

IV. Importance of Responsible Use

While tapping remarkable flexibility seems liberating, adhering to regional spectrum regulations and etiquette norms proves vital to sustaining operational privileges. We must use radio waves judiciously.

A. Radio Spectrum Management

Bands segregate specific services, while power/duty limitations prevent disruption. Licensing also designates qualified participation. Monitor usage to avoid casually wasting frequencies.

B. Permits, Licensing and Etiquette

Understand local authorization policies governing public, business, ham, and FRS bands. Seek permits only as necessary without dismissing rules casually. Always yield the right-of-way when professional services need priority access during incidents.

C. Safety and Preparedness

When making emergency plans, consider backup charging, gear protection, linking signal boosters, and fallback options in case things fail. Only risk harm to groups counting on radio access with prepping equipment to work somehow, even if everything else breaks due to the disasters. Having redundant backups makes systems more resilient when emergencies hit.

V. Wrapping UP

We covered immense ground exploring Baofeng's flexibilities, transforming simple walkie-talkies into exceptionally versatile communications solutions. Let's recap achievements powering effective capability leaps:

A. Core Competencies Gained

Readers progressed from basic radio operation familiarity to niche applications expertise - configuring encrypted sensor data backhauls, crafting disciplined emergency protocols, pairing AI assistance apps, and more.

B. Final Thoughts

Baofeng radios pack tons of ability into tiny, affordable packages. However, entirely making use requires an equal drive to learn all their features and how to expand with accessories creatively.

These little walkies gain vast possibilities with regular skills practice over time. Their low price makes access easy, but your willingness to explore makes them genuinely empowering.

Hopefully, this last chapter was a worthy wrap-up revisiting key radio lessons and celebrating how much you progressed from concept learning into confidently deploying fundamental radio skills. Retaining just 10% of core insights gained will be more helpful than starting clueless. This knowledge enormously improves technical and mental capability in dealing with future communication jams or emergency coordination needs.

As technologies progress aggressively, the core principles conveyed here maintain relevance, bridging practical use gaps over long periods. The radio fundamentals taught stand durable despite gadget fads fading quickly.

So, within these guidebook pages, you'll find assurance that you've gained versatile broadcast tools that are usable in almost all situations over the years. More significant disasters or more advanced radio systems may arise someday, but you've gained flexible skills for adapting even entry-level Baofeng foundations to meet unpredictable demands creatively.

Now you're good at configuring basic radio hardware plus interfacing clever data links as needed, your unleashed skills could tackle all kinds of challenges ahead.

BAOFENG RADIO GLOSSARY

1. Baofeng Hardware

Baofeng: is a famous manufacturer of low-cost, high-value portable handheld radio transceivers and accessories. It is also known for versatile digital communications equipment.

Transceiver: Radio device that can receive and transmit signals, facilitating two-way communication exchanges.

Squelch: An adjustable setting allows users to filter out unwanted background noise by only receiving signals stronger than a defined threshold.

Drop-In Charger: Charging cradle dock that Baofeng radios with removable batteries easily fit into to recharge battery packs conveniently overnight. Simply drop the radio with cells into the slot.

2. Radio Signal Concepts

UHF: Ultra High Frequency signals between 300 MHz and 3 GHz are used primarily for short-range two-way voice and low-bandwidth data communications when local radio wave propagation offers a practical advantage. Includes critical walkie-talkie frequencies.

VHF: Very High-Frequency signals between 30-300 MHz with slightly longer wavelengths, enabling signals to bend around obstacles in line-of-sight while retaining efficiently sized antennas. They are used for broadcasts, navigation systems, and mobile communications.

Simplex vs Duplex: Simplex uses a single shared frequency for back-and-forth relay-style communication. Duplex allows sending and receiving simultaneous discrete signals for continuous exchanges more akin to telephone conversations.

CTCSS/DCS: Analog sub-audible tone protocols transmitted with standard analog audio, allowing receiving stations to filter transmissions by configured code values for enhanced privacy and channel access control.

Bandwidth: The range of radio wave frequencies occupied by transmitted information-carrying signals. Larger bandwidth allows more information carriage but divides finite frequencies available among more concurrent users.

Frequency Hopping: Rapidly switching transmission across various frequency channels based on secret patterns shared only with intended communicators and revised dynamically. This provides security, making eavesdropping difficult.

3. Programming and Configuration

CHIRP Software: Popular third-party programming software for Baofeng radios that massively simplifies channel, code, and setting configurations compared to tedious manual entry through numeric keypads. Allows complete customization leveraging easy personal computer access.

Firmware: Low-level embedded software programmed into read-only memory chips that control baseline radio functionalities and expose capability interfaces to higher-layer applications and user adjustments.

Field Programming: Making channel, privacy code, and various operational setting changes to units conveniently over-the-air leveraging software tools rather than requiring manual tweaking or wired connection to program frequencies. Expedites field upgrades.

Memory Slots: The number of available channel storage slots for programming frequently used frequency pairs, descriptive names, and communication options into the radio for quick recall and later usage when changing between saved channels.

4. Accessories and Enhancements

GAGA Antenna: This aftermarket external rubber ducky replacement antenna offers better performance than the mediocre stock antennas shipped with commercial handheld radios. It is an easy screw-on upgrade for range/signal improvements.

Earpiece: This accessory allows private radio audio monitoring via earbuds for discrete communications and improved reception in noisy environments. The patch cord connects the accessory jack on the radio unit to the headphones.

Thumb Screw Microphone: External microphone with integrated speaker, volume/squelch controls, and large Push-to-Talk buttons. It allows convenient handheld use for long transmissions, avoiding fatiguing and tightly gripping the radio.

Spare Battery Pack: Carrying additional charged removable radio batteries ensures the ability to hot-swap faded cells for continuous operation across extended multi-day remote activities that lack charging access for indefinitely sustaining communications capability.

5. Operational Environments

Weatherproofing: Protective treatment applications and weather seals safeguard sensitive radio electronics from environmental hazards like blowing rain, dust ingress, or humidity through materials beneficially repelling water and resisting corrosion.

MIL-STD-810 Rating: A series of demanding US military tests evaluates product durability against harsh elements, including drops, shocks, vibrations, humidity, and extreme high/low temperatures. Devices earning certified ratings endure more abuse.

RF Shielding: Blocking radio frequency signals through partially conductive barrier coverings prevents ambient interference penetration and degrades radio performance, but it requires adequate ventilation pathways for thermal heat dissipation from tightly shielded transmitter electronics.

Faraday Cage: Fully enclosed metal mesh shielding surrounding sensitive radio gear creates a continuous conductive barrier, isolating devices from intense electromagnetic interference or solar flares and allowing ventilation openings. Hardens resilience.

6. Software and Connectivity

Packet Radio: An amateur radio methodology that utilizes small transmitted data packets conveying text or computer communications over radio links similar to early dial-up internet connections but creatively conveys data streams over voice-centric analog FM channels.

USB Cable: Universal serial bus cables allow convenient radio connectivity to personal computers for accessing programming software tools. They leverage desktop power to simplify device configuration edits that are replicated back to portable units.

ADI Encryption: Optional proprietary Advanced Digital Interpretation algorithm engine modules integrate AES 128-256 bit encryption between enabled radio units for securing voice communications through mathematical cryptography suited to public safety and business adaptation but at added expense over stock consumer models utilizing primary built-in inversion or rolling cipher encryption optionally for light privacy.

DMR: Digital Mobile Radio - open standard two-way digital radio protocols enhancing spectral efficiency, cross-vendor interoperability, and feature sets by leveraging advanced digital processing, signal sampling, and low-latency optimized timeslot access coordination.

7. Advanced Applications

SDR: Software-defined radio flexibility allows replacing radio hardware function blocks with adaptable software modules customized via open-source firmware enhancements. This allows adjusting filtering, modulation, bandwidths, and frequency hopping behavior according to software capabilities despite hardware radio constraints. This enables custom tweak advantages, creatively circumventing cost barriers while pursuing pure radio hardware enhancements equally.

IO Circuits: Auxiliary Input/Output integrated circuit pins built into programmable radio units allow interfacing external sensor devices or feeding pre-processed telemetry inputs into radio modems from scientific equipment. This conveys meaningful environmental sensor metrics to distant centralized observers lacking conventional wide-area infrastructure, enabling remote site flexibility.

Fox Hunting: Direction finding methodology leveraging carefully aimed high gain directional antennas with specialized receivers allowing radio localization by sweeping bearings and triangulating signal peaks visually indicated by S-meter graphs determining interfering illicit transmission origins or localizing distress beacons to expedite rescue dispatch efficiency—valuable search and rescue tool.

Tones: Audible sub-audible signaling tones prefixed to analog audio transmissions for conveying additional encoded signaling data supporting selective squad-based receives through matching squelch tone criteria or alerting emergency priority overrides. Extends channel access constraints conventionally.

Data Shot: Configuring radio transceivers into transmit-only data burst relay mode allows remote sensor-equipped stations to launch quickly compressed low-resolution telemetry packets to distant receiver logging aggregation sites across unsuitable distances impractical for real-time streaming transfers. Useful for sparse updates.

AVC: Automatic Volume Control - Self-adjusting receiver gain continually equalizing detected volume against ambient reception noise minimizes listener disruption transitioning signal quality variation extremes avoiding jarring perceived loudness shifts retaining necessary clarity particularly notable on fringe reception area mobile units contending signal turbulence factors. Reduces annoyance!

PTT ID: Push-to-Talk Identification—A transmitter radio protocol requiring periodic identification header frames prepended by unit call signs for legal operations, similar to aviation transponder squawks conveying transmitting node identifiable registration particulars associated with accountable transmission traced back to the radio equipment owner—a necessity for wide-area channel sharing access adhering to regulations.

RFI: Radio Frequency Interference - Disruptive electromagnetic radiation unintentionally emitted from inadequately shielded electronic devices encroaching spectral emission containment limits or intentionally radiating obfuscation noise sources actively attempting to block target receivers through aggressive thermal noise interference or specifically tuned targeted harmonics intermodulation deliberately Cacophonous. Regulatory restraints mitigate abuses per band.

8. Radio Frequency Principles

Wavelength: The physical dimensional length of a complete radio wave cycle measuring the distance between recurring identical points typically crest to crest as one Hertz frequency oscillation propagates linearly across a theoretical medium devoid of attenuation constraints. Dictates antenna sizing.

Harmonic Distortion: Nonlinear signal distortion generates interfering integer multiples of transmission frequencies outside allocated channel bandwidths due to overdriven amplification, overloading transmitter dynamics beyond linear resolution ranges, or spurious frequency multiplication intermodulation produced by external unmatched impedance loads. It degrades purity.

Standing Wave Ratio: The mathematical ratio between maximum and minimum AC voltages along fixed linear transmission feedline paths is caused by partial signal reflections at antenna mismatch. It indicates the effectiveness of impedance compatibility between radio transmitter design and connected antenna load terminating endpoint. Lower SWR improves efficiency.

Atmospheric Ducting: Long-range radio propagation enhancement occurs when temperature inversions create stacked horizontal atmospheric layers bending and channeling radio transmissions following duct shape geometry restricting refracting signals strengths condensed sustainably beyond standard line of sight constraints. Creates distant signal ducts!

Decay Rate: Rate quantifying rapid radio reception field intensity signal fading parameterized over increasing receiver distance or time metrics as broadcast electromagnetic waves spread inversely squared, eventually attenuating below discernible background noise thresholds combined with interceding signal degradation influences from diffraction, reflection, refraction, and obstruction factors impacting propagation. Defines rage limits ultimately.

9. Digital & Analog Protocols

P25: APCO Project 25 is an open standard suite of digital radio protocols mandated for public safety first responder interoperability, enhancing portable radio flexibility, guaranteeing vendor egalitarian data interchange, bulletproofing communications viability, evaluating extreme emergency scenarios, facilitating system convergence transitioning aging analog predecessors toward future-proof digital radio governance maximizing spectrum efficiencies invariably supporting encryption fortifying prudent security virtues underpinning mission-critical global governmental alliances twenty-five successive years henceforth!

DMR: Digital Mobile Radio - open standard two-way digital radio protocols enhancing spectral efficiency, cross-vendor interoperability, and feature sets by leveraging advanced digital processing, signal sampling, and low-latency optimized timeslot access coordination. Promotes upgrades.

NXDN: Next-generation Digital Narrowband network - proprietary over-the-air IP-linked digital radio signaling protocol consolidating 6.25 kHz FDMA channel pairs under unified coordinates fostering systemwide resource efficiencies, enhanced security measures, lower power consumption with dual-mode gateway transitions between legacy analog cellular carriers and next-generation LTE broadband gateway telecommunications evident universally. Embraces progress.

10. Hardware Installation Factors

Pigtail: A short, flexible coaxial cable jumper connecting radio units to external antenna suspensions. Dedicated 50-ohm impedance RG-58 or mini-UHF varieties are typically available, terminated with male/female chassis connectors on opposing ends. This allows relocation options away from units, extending antenna positioning liberties remotely.

Counterpoise: Parallel ground-plane wires or mesh grid conductors elevated above the earth's natural polarity to enhance vertical antenna gain performance figures by symmetrically balancing induced signal currents collected improving baseline efficiency parameters while securing elevated isolation from detuning earth ground losses beneficially mitigating electromagnetic absorption, improving resonating efficiency dramatically given proper installation methodologies understood.

ATU: Antenna Tuning Unit - Interposed matching network circuit between a radio transmitter and target antenna structure attempting impedance matching maximization through manipulatable capacitance and inductance topographical refinements strategically adjusting resonating reactance according toward ideal 50-ohm conjugate impedance matching scenario mitigating standing wave deficiencies reflected from suboptimal antenna impedance mismatches all too common lacking appropriate tuning remediation attentiveness degrading efficient range propagation opportunities unnecessarily. Rescue investments!

11. Advanced Radio Techniques

I/Q Sampling: Radio-defined software modulation-demodulation methodology numerically captures analog signal set input representations across orthogonal dual vectors denoted in-phase (I) and quadrature-phase (Q) components. This facilitates accurate analog preservation within minimal digital sampling datasets for subsequent manipulation, encoding/decoding, or frequency conversion before analog regeneration. It enables software radio flexibility.

Direct Sampling: Innovative Software Defined Radio receiver architecture eschewing traditional preceding heterodyning frequency conversion stages while directly subsampling desired spectrum band segments leveraging high-speed digital sampling synthesizing final demodulated reception product purely computationally. Reduces costs!

ONE LAST THING

Dear Radio Enthusiast,

As we tune in to the final frequency of this radio guide, I want to extend my heartfelt gratitude for embarking on this journey with me. It has been an absolute pleasure to share my knowledge and passion for Baofeng radios with you, and I sincerely hope that this guide has significantly enhanced your radio experiences.

One of the most gratifying ways to give back to the radio community is by leaving a review for this guide. Your insights, experiences, and suggestions could greatly assist fellow radio enthusiasts in making informed decisions, whether it's choosing the right model, troubleshooting common issues, or exploring new techniques.

Dear reader, I humbly request your assistance in the form of a review. Your words have the power to inspire others to delve deeper into the world of Baofeng radios and experience the joy of connecting through the airwaves. Your honest feedback is invaluable to me as an author, as it helps me understand what resonated with you, what areas may need improvement, and how I can better tailor future resources to meet the needs of the radio community.

With warmest regards,

Patrick Vincent.

**You can leave a review
by scanning this QR Code**